CLASSICAL
ASPHERICAL MANIFOLDS

Conference Board of the Mathematical Sciences
REGIONAL CONFERENCE SERIES IN MATHEMATICS

supported by the
National Science Foundation

Number 75

CLASSICAL ASPHERICAL MANIFOLDS

F. Thomas Farrell
L. Edwin Jones

Published for the
Conference Board of the Mathematical Sciences
by the
American Mathematical Society
Providence, Rhode Island

Expository Lectures
from the CBMS Regional Conference
held at the University of Florida
January 9–14, 1989

Research supported by National Science Foundation Grants DMS-88-01312 and
DMS 88-03056.

1980 *Mathematics Subject Classification* (1985 *Revision*). Primary 18F25, 22E40,
53C20, 57R55.

Library of Congress Cataloging-in-Publication Data
Farrell, F. Thomas, 1941–
    Classical aspherical manifolds/F. Thomas Farrell, L. Edwin Jones.
    p. cm. —(Regional conference series in mathematics, ISSN 0160-7642; v. 75)
    "Expository Lectures from the CBMS Regional Conference Series held at the
University of Florida, January 9–14, 1989."
    Includes bibliographical references.
    ISBN 0-8218-0726-9 (alk. paper)
    1. Manifolds (Mathematics)–Congresses. I. Jones, L. Edwin, 1945–.
II. Conference Board of the Mathematical Sciences. III. National Science
Foundation (U.S.). IV. Title. V. Series.
QA1.R33 no. 75
[QA613]
510 s—dc20
[514′.223]

90-39

CIP

10 9 8 7 6 5 4 3 2 1   95 94 93 92 91 90

Contents

Introduction

An NSF-CBMS regional conference on "K-theory and Dynamics" was held in Gainesville, Florida from January 9th to the 14th, 1989. The conference was funded by the National Science Foundation and hosted by the University of Florida. C. W. Stark was the conference director and the organizing committee consisted of D. Fried, L. E. Jones, and J. B. Wagoner. F. T. Farrell was the principal lecturer and this book is based on those lectures. The more technical topics have been deleted. Some results which were obtained since the conference are discussed in an epilogue (Chapter 6). This epilogue also contains some theorems, not mentioned at the conference but obtained in the last ten years, showing there are many aspherical manifolds which are not classical. The ten lectures were primarily concerned with classical aspherical manifolds; e.g., those arising as double coset spaces of Lie groups or from synthetic geometry. The main problem addressed was the topological characterization of compact (closed) classical aspherical manifolds. The problem has been mostly solved; 3-dimensional and 4-dimensional manifolds present the most important unsolved aspects. (Poincaré's conjecture is closely related to the 3-dimensional problem.)

We wish to express our special thanks to Chris Stark. His efforts made possible both the conference and this book.

F. Thomas Farrell
L. Edwin Jones

1. The Structure of Manifolds from a Historical Perspective

The classification of manifolds has been an enduring desire of topologists. They attempt to count up to homeomorphism (or diffeomorphism) all manifolds homotopically equivalent to a given topological space X. We proceed to formulate this program more precisely. A pair (M, f) is a homotopy-topological structure on X provided $f: M \to X$ is a homotopy equivalence and M is a closed manifold. (A closed manifold is a compact manifold without boundary.) If M is a smooth manifold, then (M, f) is called a homotopy-smooth structure on X. Two homotopy-topological structures (M_i, f_i) on X, where $i = 1$, or 2, are equivalent when there exists a homeomorphism $h: M_1 \to M_2$ such that the composite $f_2 \circ h$ is homotopic to f_1. Smooth structures are equivalent if, in addition, h is a diffeomorphism.

1.1. DEFINITION. The set of equivalence classes of homotopy-topological structures on X is denoted $\mathscr{S}(X)$ and equivalence classes of homotopy-smooth structures is denoted $\mathscr{S}^{smooth}(X)$.

1.2. FUNDAMENTAL PROBLEM. Calculate the cardinalities of $\mathscr{S}(X)$ and $\mathscr{S}^{smooth}(X)$. For example, when are they nonempty? When are they finite sets?

When X is itself a closed manifold whose dimension is either 1 or 2, this problem is solved and the answer is $|\mathscr{S}(X)| = |\mathscr{S}^{smooth}(X)| = 1$. This is because we can explicitly list the closed 1 and 2 dimensional manifolds. The circle S^1 is the only closed one-dimensional manifold. The sphere, projective plane, torus, and finite connected sums of these constitute all the closed two-dimensional manifolds. On the other hand, one of the oldest unsolved problems in topology is the special case of 1.2, formulated by Poincaré [1904], where X is the three-dimensional sphere S^3.

1.3. POINCARÉ'S CONJECTURE. $|\mathscr{S}(S^3)| = 1$.

1.4. GENERALIZED POINCARÉ CONJECTURE. $|\mathscr{S}(S^n)| = 1$, where S^n denote the n-dimensional sphere.

REMARKS. Conjecture 1.4 has been shown to be true except in the original case of 1.3; i.e., except when $n = 3$. For $n > 5$, Smale [1961] showed that the image of the obvious map from $\mathscr{S}^{smooth}(S^n)$ into $\mathscr{S}(S^n)$ contains only one element. Stallings [1965] extended Smale's result to include $n = 5$

1

and Newman [**1966**] completed the verification of 1.4, provided $n \neq 3, 4$. Freedman [**1982**] showed 1.4 is true when $n = 4$.

The solution of 1.4 $(n \neq 3)$ depends on an analysis of h-cobordisms.

1.5. DEFINITION. A compact (connected) manifold W whose boundary ∂W is the disjoint union of two closed manifolds N^- and N^+ is called an h-cobordism, provided the inclusions of N^- into W and of N^+ into W are both homotopy equivalences. The pair (W, N^-) is called a h-cobordism with base N^-, and top N^+. A smooth h-cobordism is one where W is a smooth manifold.

Note that an h-cobordism with base N is homotopically equivalent to the cylinder $N \times [0, 1]$, which is called the trivial or product h-cobordism with base N. Two topological (smooth) h-cobordisms W_1 and W_2 with the same base N are equivalent provided there is a homeomorphism (diffeomorphism) from W_1 to W_2 which is the identity map when restricted to N.

1.6. h-COBORDISM THEOREM. *Any simply connected h-cobordism W is a product provided* $\dim W \neq 3$ *and* 4.

REMARKS. Smale [**1962**] showed W is diffeomorphic to a cylinder provided W is smooth and $\dim W > 5$. But Donaldson [**1983**] showed the analogue of Smale's result is false when $\dim W = 5$. Kirby and Siebenmann [**1977**] proved 1.6 when $\dim W > 5$, and Freedman [**1982**] proved 1.6 when $\dim W = 5$. Stallings [**1962**] showed 1.6, when $\dim W = 3$, is equivalent to 1.3.

We now show how to deduce, 1.4, when $n > 4$, from the h-cobordism theorem. Let Σ be a closed manifold which is homotopically equivalent to S^n and let B^- and B^+ be a pair of disjoint (closed) n-dimensional balls nicely embedded in Σ. Set W equal to Σ with the interiors of B^- and B^+ deleted. Then W is a simply connected h-cobordism whose base is S^{n-1}. Because of 1.6, W is homeomorphic to $S^{n-1} \times [0, 1]$. By placing the balls B^+ and B^- back onto the top and bottom of this cylinder, respectively, we see that Σ is homeomorphic to S^n. This last step requires an important observation due to Alexander [**1923**], which is sometimes called *Alexander's Trick*.

1.7. THEOREM. *Any homeomorphism of the n-sphere extends to a homeomorphism of its bounding n + 1-ball.*

PROOF. Let $\mathbb{B}^{n+1}$ denote $\{x \in \mathbb{R}^{n+1} \mid |x| \leq 1\}$ and S^n denote $\{x \in \mathbb{B}^{n+1} \mid |x| = 1\}$. To each continuous function $f: S^n \to S^n$, associate its "cone" $\overline{f}: \mathbb{B}^{n+1} \to \mathbb{B}^{n+1}$ defined by

$$\overline{f}(x) = \begin{cases} \dfrac{1}{|x|} f(x) & \text{if } x \neq 0 \\ 0 & \text{if } x = 0. \end{cases}$$

Clearly, $\overline{f}$ extends f. Furthermore, if f is a homeomorphism, so is $\overline{f}$. $\quad\square$

REMARKS. When f is a diffeomorphism, $\bar{f}$ need not be differentiable at 0. In proving 1.4 (when $n = 4$), a h-cobordism W is constructed between Σ and S^4 using surgery theory. Since W is homeomorphic to $S^4 \times [0, 1]$ by 1.6, S^4 is homeomorphic to Σ.

Let us now look at the smooth analogue of 1.4 (when $n > 5$); i.e., we wish to calculate $|\mathscr{S}^{smooth}(S^n)|$. Let Σ be a closed smooth manifold which is homotopically equivalent to S^n and let W be Σ with the interiors of two disjoint, smoothly embedded (closed) n-balls deleted. By the smooth h-cobordism theorem of Smale [1962], W is diffeomorphic to $S^{n-1} \times [0, 1]$. Hence Σ is diffeomorphic to this cylinder with balls attached smoothly to its top and bottom. If 1.7 were true when the word "homeomorphism" is replaced by "diffeomorphism", then $|\mathscr{S}^{smooth}(S^n)| = 1$, provided $n > 5$. But Milnor [1956] showed this is false; in particular $|\mathscr{S}^{smooth}(S^7)| > 1$. Consequently, Alexander's Trick fails in the smooth category; in fact, $\mathscr{S}^{smooth}(S^n)$ measures this failure when $n > 5$. Kervaire and Milnor [1963] developed a theory for calculating $\mathscr{S}^{smooth}(S^n)$ when $n > 4$. One consequence of this theory is the following finiteness result.

1.8. THEOREM. When $n > 4$, $\mathscr{S}^{smooth}(S^n)$ is a finite set.

Forty years ago, the available techniques were so limited that the following vast generalization of 1.4, attributed by Novikov [1965] to Hurewicz, seemed possible.

1.9. HUREWICZ'S PROBLEM. Are homotopically equivalent, simply connected, closed manifolds always homeomorphic?

But Novikov [1965] showed the answer to this question is no. The following is an outline of how this is done. Browder [1962] and Novikov [1964] independently developed a theory to analyze $\mathscr{S}^{smooth}(X)$, when X is simply connected. It built on Kervaire and Milnor [1963] and Smale [1962] and is called (simply-connected) surgery theory. One consequence, due to Novikov [1964], is the following result.

1.10. THEOREM. Let M be a closed, simply connected, smooth manifold with $\dim M > 4$. Then $|\mathscr{S}^{smooth}(M)| = \infty$ if and only if there exists a positive integer n, with $4n < \dim M$, such that $H_{4n}(M, \mathbb{Q}) \neq 0$.

Recall that the rational Pontryagin classes $p_n M \in H^{4n}(M, \mathbb{Q})$ are invariants of diffeomorphisms $f: N \to M$ between smooth manifolds; i.e., $f^*(p_n M) = p_n N$. In proving 1.10, Novikov [1964] showed that if $|\mathscr{S}^{smooth}(M)| = \infty$, then there exists a homotopy-smooth structure (N, f) on M such that $f^*(p_n M) \neq p_n N$, for some integer n. Applying this to $M = S^4 \times S^5$, shows there exists a closed smooth manifold N homotopically equivalent to $S^4 \times S^5$ with $p_1 N \neq 0$. Since $p_1(S^4 \times S^5) = 0$, $S^4 \times S^5$ and N are not diffeomorphic. Thus the answer to 1.9, in the smooth category is no. The manifold N can be explicitly constructed, independent of Novikov [1964], as the total space of the sphere bundle associated to a six-dimensional real vector bundle η with base space S^4. Under the normal bijective correspondence of such bundles with the ∞-cyclic group $\pi_3(SO(6))$, η can be 24 times a generator; cf. Milnor and Stasheff [1974, p. 245].

Novikov [1966] proved the following foundational result in the topological theory of manifolds.

1.11. THEOREM. *The rational Pontryagin classes are topological invariants; i.e., if $f: N \to M$ is a homeomorphism between smooth manifolds, then* $f^*(p_n M) = p_n N$.

Consequently, the manifold N, constructed above, cannot be homeomorphic to $S^4 \times S^5$ giving no as the answer to 1.9. (In Novikov's original argument, $M = S^4 \times S^2$ and he uses an earlier partial result on 1.11 proved in Novikov [1965].)

Besides tangent bundle invariants, like Pontryagin classes, there is an extra invariant which plays an important role in understanding 1.2 when X is not simply connected. To any homotopy equivalence $f: Y \to X$ between finite simplicial complexes, Whitehead [1939] associated an element $\tau(f)$ in an abelian group $\mathrm{Wh}(\pi_1 X)$; $\tau(f)$ is called the Whitehead torsion of f and $\mathrm{Wh}(\pi_1 X)$ the Whitehead group of $\pi_1 X$; cf. Cohen [1973] for a detailed discussion of these. In fact, given any ring R with identity, an abelian group $K_1 R$ is defined by abelianizing the general linear group of R. More precisely,

$$K_1 R = \lim_{n \to \infty} \mathrm{GL}_n(R)/[\mathrm{GL}_n(R), \mathrm{GL}_n(R)]$$

where $\mathrm{GL}_n R$ maps to $\mathrm{GL}_{n+1} R$ by sending a $n \times n$ matrix A to the $(n+1) \times (n+1)$ matrix

$$\begin{pmatrix} A & 0 \\ 0 & 1 \end{pmatrix}.$$

Let $\mathbb{Z}\Gamma$ denote the integral group ring of the group Γ. (Recall $\mathbb{Z}\Gamma$ is the ring of all $\mathbb{Z}$ valued functions on Γ having finite support. Addition is pointwise and convolution is the multiplication.) Each element γ in Γ determines elements γ and $-\gamma$ in $\mathrm{GL}_1(\mathbb{Z}\Gamma)$ and hence in $K_1 \mathbb{Z}\Gamma$. (Identity γ with the function $\Gamma \to \mathbb{Z}$ whose support is γ and whose value is 1.) Let $\{\pm\gamma | \gamma \in \Gamma\}$ denote the subgroup of $K_1 \mathbb{Z}\Gamma$ formed by all these elements.

1.12. DEFINITION. $\mathrm{Wh}\,\Gamma = K_1 \mathbb{Z}\Gamma / \{\pm\gamma | \gamma \in \Gamma\}$.

The Whitehead group is a functor from groups to abelian groups. Homotopic maps have the same Whitehead torsion and Chapman [1974] proved the following basic fact.

1.13. THEOREM. *If f is a homeomorphism, then* $\tau(f) = 0$.

Whitehead torsion is particularly useful in classifying h-cobordisms W with a given (nonsimply connected) base N. An element $\tau(W, N) \in \mathrm{Wh}\,\pi_1 N$

is defined by $\tau(W, N) = \tau(r)$ where $r: W \to N$ is any retraction. (Kirby and Siebenman [1977] showed, before 1.13 was proven, that $\tau(W, N)$ is defined even if N or W cannot be triangulated.) This leads to the s-cobordism theorem due (in the smooth category) to Barden [1963], Mazur [1963] and Stallings [1965]. It was extended by Kirby and Siebenman [1977] to the topological category.

1.14. s-COBORDISM THEOREM. *There is a bijection, given by $\tau(W, N)$, between the set of homeomorphism classes of h-cobordisms (W, N) with a given base N and the set $\mathrm{Wh}\,\pi_1 N$, provided $\dim N > 4$. (If N is a smooth manifold, then $\tau(W, N)$ is also a bijection between diffeomorphism classes of smooth h-cobordisms and $\mathrm{Wh}\,\pi_1 N$.) The cylinder corresponds to 0 under this bijection.*

REMARK. We sometimes abbreviate $\tau(W, N)$ to $\tau(W)$ provided it is unambiguously understood which component of ∂W is the base N of W.

The ring $\mathbb{Z}\Gamma$ is equipped with an antiautomorphism (conjugation) defined by $\overline{f}(\gamma) = f(\gamma^{-1})$, where $f \in \mathbb{Z}\Gamma$ and $\gamma \in \Gamma$. It determines an involution $x \mapsto x^*$ of $\mathrm{Wh}\,\Gamma$ induced by sending an element in $\mathrm{GL}_n(\mathbb{Z}\Gamma)$ to its conjugate transpose matrix. Stallings [1965] proved the following result.

1.15. THEOREM. *Let W be an h-cobordism with base N^- and top N^+, then*

$$\tau(W, N^+) = (-1)^n \tau(W, N^-)^*$$

where $n = \dim N^- > 4$.

REMARKS. The composite of the inclusion of N^+ into W with a retraction of W onto N^- induces an isomorphism of $\pi_1 N^+$ to $\pi_1 N^-$. In this way, $\mathrm{Wh}\,\pi_1 N^+$ is identified with $\mathrm{Wh}\,\pi_1 N^-$. This identification is implicitly used in 1.5.

Let C_n denote the cyclic group of order n. The following result is a consequence of 1.13, 1.14 and 1.15 together with a calculation due to Higman [1940]

1.16. EXAMPLE. Let M be any odd dimensional, closed manifold with $\pi_1 M = C_5$ and $\dim M > 4$; e.g., M can be a lens space. Then $|\mathscr{S}(M)| = \infty$.

PROOF. There is associated to each element x in $\mathrm{Wh}\,C_5$ a homotopy-topological structure (N_x, f_x) on M as follows. The manifold N_x is the top of the h-cobordism W_x with base M such that $\tau(W_x, M) = x$. The map f_x is the composite of the inclusion $\sigma_x: N_x \to W_x$ with a retraction $r_x: W_x \to M$. The Whitehead torsion of a composite is the sum of their torsions. Consequently, we compute $\tau(f_x)$ as follows:

$$\tau(f_x) = \tau(r_x) + \tau(\sigma_x) = x - (-1)^n x^*$$

where $n = \dim M$. Higman [1940] showed that $\mathrm{Wh}\,C_5$ contains an element

y of infinite order, with $y^* = y$. (If t is a generator of C_5, then y is represented by the element $t + t^{-1} - 1$ in $GL_1(\mathbb{Z}\Gamma)$.) Since $\tau(f_{ny}) = 2ny$, it follows from 1.13, that the structures (N_{ny}, f_{ny}), where $n \in \mathbb{Z}$, are pairwise inequivalent. □

REMARKS. See Cohen [1973] for more details; in particular, the conclusion of 1.16 holds under the more general assumption that $\pi_1 M = C_p$, where p is any prime ≥ 5.

So it seems that either too much homology in M or too much fundamental group causes $|\mathscr{S}(M)| > 1$. But there is some cancellation among these two. Hirzebruch [1966] showed there is a constraint on the rational Pontryagin classes p_n of homotopically equivalent manifolds. There are classes $L_n M \in H^{4n}(M, \mathbb{Q})$ associated to any manifold M such that $L_n M$ is a (universal) polynomial in $p_1 M$, $p_2 M$, ..., $p_n M$ and $p_n M$ is a (universal) polynomial in $L_1 M$, $L_2 M$, ..., $L_n M$. In light of Kirby–Siebenmann [1977], Hirzebruch's result has the following implication.

1.17. HIRZEBRUCH INDEX THEOREM. *Let* (N, f) *be a homotopy-topological structure on a closed orientable manifold* M. *Then* $f^*(L_n M) = L_n N$, *provided* $\dim M = 4n$.

Hirzebruch proved 1.17 by identifying the number $\langle L_n M, [M] \rangle$ with the index of M, which is defined in terms of the cohomology ring of M. (Here $\langle \, , \, \rangle$ denotes the Kronecker product and $[M]$ an orientation class for M.) In the same paper where he answered 1.9 negatively, Novikov [1964] extended 1.17 by showing it is also true when $\dim M = 4n+1$. Note that this extension is interesting only when $\pi_1 M \neq 1$, since otherwise $H^{4n}(M, \mathbb{Q}) = 0$ by Poincaré duality. Novikov noted that $H^1(M) = [M, S^1]$; i.e., the group of homotopy classes of maps of M into S^1. He then showed geometrically that the number $\langle L_n M \cup a, [M] \rangle$, where $a \in H^1(M)$, can be interpreted as a type of higher index of the ∞-cyclic cover of M associated to a. Moreover, he conjectured that $\langle L_n M \cup a_1 \cup a_2 \cup \cdots \cup a_s, [M] \rangle$ is also a homotopy invariant for any set of one-dimensional (integral) cohomology classes $a_1, a_2, \ldots, a_s$ in $H^1(M)$, when $\dim M = 4n + s$. This conjecture was verified by Rohlin [1966] when $s = 2$. It was later verified for arbitrary s by Farrell and Hsiang, cf. Hsiang [1969], and independently by Kasparov [1970]. Both arguments used the results about fibering manifolds over a circle contained in Farrell [1971]. Novikov eventually generalized his conjecture to the following.

1.18. NOVIKOV'S CONJECTURE. *Let* (N, f) *be a homotopy-topological structure on a closed manifold* M, *where* $\dim M = 4n + k$. *Then the cohomology class* $L_n N - f^*(L_n M)$ *is annihilated by cup product with any class in* $H^k(N, \mathbb{Q})$ *which is in the image of* ϕ^*, *where* $\phi \colon N \to K(\pi_1 N, 1)$ *classifies* $\pi_1 N$.

REMARKS. This unsolved conjecture is the center of much current interest. There have been many important partial results. See Kasparov [**1988**] for its status as of 1987.

1.19. DEFINITION. A connected (locally contractible) topological space X is aspherical provided $\pi_n X = 0$ for all $n > 1$; i.e., provided its universal cover is contractible. Such a space is an Eilenberg–Mac Lane space $K(\pi_1 X, 1)$.

Closed aspherical manifolds occur in many contexts. The following are some important examples. The circle S^1 is the simplest one and another basic example is the n-dimensional torus $S^1 \times S^1 \times \cdots \times S^1$ (n-factors) which is constructed from it. Low-dimensional topology is a source of many examples. All closed two-dimensional manifolds, except the sphere and projective plane, as well as all closed, irreducible three-dimensional manifolds with infinite fundamental groups are examples.

Examples also arise in Lie group theory; namely, all double coset spaces $\Gamma \backslash G / K$ are closed aspherical manifolds where G is a (virtually connected) Lie group, Γ is a torsion-free discrete cocompact subgroup of G, and K is a maximal compact subgroup of G. (Given any property $\mathscr{P}$, such as being connected, a group is *virtually* $\mathscr{P}$ provided it contains a subgroup of finite index which has property $\mathscr{P}$.) The universal cover of $\Gamma \backslash G / K$ is G/K which is diffeomorphic to $\mathbb{R}^n$ for some integer n. These examples include compact infranilmanifolds and compact infrasolvmanifolds. (These manifolds can be defined as double coset spaces $\Gamma \backslash G / K$ where G is either virtually nilpotent or solvable, respectively.)

Finally, Riemannian geometry is a rich source of examples. Any compact (connected) nonpositively curved Riemannian manifold M^n is aspherical since its universal cover is diffeomorphic to $\mathbb{R}^n$. (A Riemannian manifold is flat, negatively, nonpositively, positively, nonnegatively curved provided all its sectional curvature values are zero, negative, nonpositive, positive, nonnegative, respectively.)

Both the Lie group and the Riemannian manifold examples include all the compact (connected) flat or hyperbolic manifolds and, more generally, all compact (connected) nonpositively curved locally symmetric spaces. (Recall a Riemannian manifold is a locally symmetric space provided its sectional curvatures stay constant during parallel translation.) But there are aspherical manifolds in each class which are not in the other.

REMARK. If Novikov's conjecture were true, then his method for constructing a negative solution to Hurewicz's problem (by finding homotopically equivalent manifolds with different rational Pontryagin classes) would fail for any aspherical manifold M; i.e., would not show $|\mathscr{S}(M)| > 1$.

1.20. DEFINITION. A group Γ is *torsion-free* provided the only element of finite order in Γ is its identity element.

REMARK. The fundamental group of an aspherical manifold is torsion-free since a $K(C_n, 1)$ cannot be finite dimensional when $n > 1$.

1.21.   WHITEHEAD TORSION CONJECTURE. If Γ is torsion-free, then $\text{Wh}\,\Gamma = 0$.

REMARK. If 1.21 is true, then the method of Example 1.16 for showing $|\mathscr{S}(M)| > 1$ would also fail when M is an aspherical manifold.

The following alternative to Hurewicz's problem was proposed by Borel in 1952.

1.22.   BOREL'S CONJECTURE. If M is a closed aspherical manifold, then $|\mathscr{S}(M)| = 1$.

REMARK. As was shown above, 1.22 fits well with both Novikov's conjecture and the Whitehead torsion conjecture. Also it implies Poincaré's conjecture. This is seen as follows. Suppose Σ is closed manifold homotopically equivalent, but not homeomorphic, to S^3. Let M be any closed aspherical three-manifold; e.g., $M = S^1 \times S^1 \times S^1$. Then the connected sum $M\#\Sigma$ is homotopically equivalent, but not homeomorphic to M, because of Milnor [1962].

Borel was led to his conjecture by the results of Bieberbach [1912] and Mostow [1954].

1.23.   BIEBERBACH'S RIGIDITY THEOREM. *Let $f: N \to M$ be a homotopy equivalence between compact (connected) flat Riemannian manifolds, then f is homotopic to an affine diffeomorphism. (A map is affine provided it sends geodesics to geodesics.)*

Mostow [1954] showed that compact solvmanifolds with isomorphic fundamental groups are diffeomorphic.

REMARK. Since an aspherical manifold is a $K(\pi, 1)$-space, aspherical manifolds are homotopically equivalent provided their fundamental groups are isomorphic.

While these results of Bieberbach and Mostow make 1.22 credible, they do not verify that $|\mathscr{S}(M)| = 1$ when M is a flat Riemannian manifold or when M is a solvmanifold. (Otherwise, Poincaré's conjecture would be true since $S^1 \times S^1 \times S^1$ is both a flat Riemannian manifold and a solvmanifold.) The missing ingredient is the following. If (N, f) is a homotopy-topological structure on M, and M has some added structure (for instance, M is Riemannian flat or a solvmanifold) there is no a priori way of equipping N with this extra structure.

REMARK. Conjecture 1.22 is false in the smooth category. Browder [1965] showed that $|\mathscr{S}^{smooth}(T^7)| > 1$, where T^n is the n-dimensional torus; i.e., $T^n = S^1 \times S^1 \times \cdots S^1$ (n factors). This is a consequence of smoothing theory and the fact that the suspension of T^n is homotopically equivalent to a wedge of spheres. In fact, if Σ is any closed smooth manifold homeomorphic, but not diffeomorphic to S^n (where $n > 4$) then the connected sum $T^n\#\Sigma$ is homeomorphic, but not diffeomorphic, to T^n.

1.24. DEFINITION. A connected complete Riemannian manifold is called *hyperbolic* or *spherical* if its sectional curvatures are identically -1 or $+1$, respectively.

As we have already noted, closed hyperbolic manifolds are aspherical. On the other hand, closed spherical manifolds are not aspherical since their universal covers are spheres; cf., Hicks [**1971, p. 168**]. Mostow [**1967**] proved the following result which gives additional support to 1.22.

1.25. MOSTOW'S RIGIDITY THEOREM. (*first case*). *Let* $f: N \to M$ *be a homotopy equivalence between closed hyperbolic manifolds, then* f *is homotopic to a diffeomorphism. In fact, if* $\dim M > 2$, *then there is a unique isometry homotopic to* f.

REMARK. Spherical manifolds are not rigid. Lens spaces are spherical manifolds and there are many examples of homotopically equivalent lens spaces which are not homeomorphic. The three-dimensional lens spaces $L_{7,1}$ and $L_{7,2}$ are examples of this; cf., Cohen [**1973, p. 98**]. Although $L_{7,1}$ and $L_{7,2}$ are homotopically equivalent, it can be shown that $\tau(f) \neq 0$ for any homotopy equivalence $f: L_{7,1} \to L_{7,2}$. Hence 1.13 shows that f cannot be a homeomorphism. This example was actually known before 1.13 was proved; namely, after Moise [**1952**] proved the Hauptvermutung for three-dimensional manifolds. This example is the reason manifolds are simply connected in Hurewicz's problem.

Further evidence supporting 1.22 is given in Mostow [**1973**] where he generalizes 1.25.

1.26. MOSTOW'S RIGIDITY THEOREM. *Let* $f: N \to M$ *be a homotopy equivalence between compact, non-positively curved, locally symmetric spaces. Assume that* M *has no closed one or two dimensional geodesic subspaces which are direct factors locally. Then* f *is homotopic to a diffeomorphism, which becomes an isometry after the Riemannian metric on* M *is changed by adjusting its normalizing constants.*

We end this chapter by discussing a result which casts some doubt on both 1.21 and 1.22. The first case of this result is due to Berridge and Dunwoody [**1979**] and the general result is in Artamonov [**1981**].

1.27. THEOREM. *Let* Γ *be the fundamental group of any closed flat Riemannian manifold with cyclic holonomy but excluding tori. (That is,* Γ *is any torsion-free, nonabelian extension of a finite cyclic group by a finitely generated abelian group.) Then there exists a finitely generated projective* $\mathbb{Z}\Gamma$ *module which is not free.*

REMARKS. The holonomy group of a flat Riemannian manifold consists of all isometries of a tangent space obtained by parallel translation around a loop. For example, the holonomy group of a flat Klein bottle is cyclic of order 2. The solution of Serre's Conjecture by Quillen [1976] and Suslin [1976] has been extended by Swan [1978] to yield that every finitely generated projective $\mathbb{Z}\Gamma$ module is free when Γ is a finitely generated free abelian group.

We proceed to show why 1.27 is disturbing. Related to K_1 are two other functors $\tilde{K}_0$ and Nil from rings with 1 to abelian groups. The projective class group $\tilde{K}_0 R$ of a ring R is the additive group with one generator (P) for each finitely generated projective R-module P and it has the following relations. If P is free, then $(P) = 0$. And for any pair of finitely generated projective modules P and Q over R,

$$(P \oplus Q) = (P) + (Q).$$

1.28. DEFINITION. A finitely generated projective module P is *stably free* if there exists a finitely generated free module F such that $P \oplus F$ is free.

REMARKS. It can be shown that any element in $\tilde{K}_0 R$ can be represented as the class of a finitely generated projective R-module P, and that $(P) = 0$ if and only if P is stably free.

The additive group Nil R has one generator (N) for each $(n \times n)$ nilpotent matrix N with entries in R (where n is any positive integer). It has the following relations in which N, N_0, N_1 are nilpotent matrices and A is an invertible matrix with entries in R:

(i) $(0) = 0$,

(ii) $(N) = (ANA^{-1})$,

(iii) $\left(\begin{bmatrix} N_0 & X \\ 0 & N_1 \end{bmatrix} \right) = (N_0) + (N_1)$.

The matrix X in (iii) is arbitrary except its entries must be from R and it must have the same number of rows as N_0 and columns as N_1.

In the following formula from Bass, Heller, Swan [1964], T denotes the ∞-cyclic group.

1.29. THEOREM. *Let Γ be any group, then* $\mathrm{Wh}(\Gamma \times T)$ *is isomorphic to the direct sum*

$$\mathrm{Wh}\,\Gamma \oplus \mathrm{Nil}\,\mathbb{Z}\Gamma \oplus \mathrm{Nil}\,\mathbb{Z}\Gamma \oplus \tilde{K}_0\mathbb{Z}\Gamma.$$

Since $\Gamma \times T$ is torsion-free whenever Γ is torsion-free, the truth of 1.21 would force the following to be true.

1.30. PROJECTIVE CLASS GROUP CONJECTURE. Let Γ be a torsion-free group, then every finitely generated projective $\mathbb{Z}\Gamma$-module is stably free.

REMARK. The truth of 1.22 would force 1.21 to be true at least when Γ is the fundamental group of any closed aspherical manifold M. This is seen as follows. Suppose x is a non-zero element in $\mathrm{Wh}\,\Gamma$ and let (W, M) be an h-cobordism whose torsion is x. Let N be the top of W and $f: N \to M$ be the composite of the inclusion map of N into W with a retraction of W onto M. If (N, f) is equivalent to (M, id) in $\mathscr{S}(M)$, then we can assume that f is a homeomorphism. Let W_f be the manifold obtained as the quotient space of W by identifying M to N via f. There is a naturally induced homotopy equivalence $\overline{f}: W_f \to M \times S^1$ and $\tau(\overline{f}) = \sigma_*(x)$ where $\sigma: \Gamma \to \Gamma \times T$ is inclusion into the first factor. Since projection of $\Gamma \times T$ onto Γ composed with σ is the identity map, $\sigma_*(x) \neq 0$. Consequently, $(W_f, \overline{f})$ and $(M \times S^1, \mathrm{id})$ are distinct elements of $\mathscr{I}(M \times S^1)$ because of 1.13.

But any nonfree projective module in 1.27 must be stably free. (This is shown in Chapter 2.) Hence 1.27 does not give counterexamples to conjectures 1.21, 1.22 and 1.30. On the other hand, to come "so close" with some of the most basic aspherical manifolds is puzzling and should be better understood.

2. Flat Riemannian Manifolds and Infrasolvmanifolds

We now restrict the term *flat Riemannian* manifold to its classical usage.

2.1. DEFINITION. A connected, complete Riemannian manifold is called *flat Riemannian* provided all its sectional curvatures are 0.

REMARK. Any noncompact, connected, parallelizable manifold supports a Riemannian metric with identically zero sectional curvatures. This is a consequence of the Smale–Hirsch codimension-zero immersion theory; cf. Hirsch [1959]. For example, let M be $S^1 \times S^2$ with a single point deleted. Since M is a parallelizable manifold, it thus supports a Riemannian metric with identically zero sectional curvatures. Note that M is not aspherical since $\pi_2 M \neq 0$. But the universal cover of a flat Riemannian manifold is isometrically equivalent to Euclidean space; cf. Hicks [1971, p. 168].

We now state and prove the result alluded to at the end of Chapter 1. It is from Farrell and Hsiang [1978].

2.2. THEOREM. *If M is a flat Riemannian manifold, then both* $\mathrm{Wh}\,\pi_1 M$ *and* $\tilde{K}_0(\mathbb{Z}\pi_1 M)$ *are zero; i.e., contain only the element 0.*

To prove 2.2, we need an algebraic structure theorem for the fundamental group of flat Riemannian manifolds. We call these groups *Bieberbach groups*. Bieberbach [1910] and Zassenhaus [1948] showed these are precisely the finitely generated, torsion-free groups Γ which are virtually abelian. More precisely there exists a unique normal abelian subgroup A with finite index in Γ such that the quotient group $G = \Gamma/A$ acts effectively on A. The group G is called the holonomy group of A and the rank of Γ is defined to be the number of ∞-cyclic summands in a direct sum decomposition of A. Recall T denotes the ∞-cyclic group.

2.3. LEMMA. *Let Γ be a Bieberbach group of rank n and holonomy group G, then either*

1. *$\Gamma = \pi \rtimes T$;*
2. *$\Gamma = B \underset{D}{*} C$ where D has index 2 in both B and C or*
3. *There exists an ∞-sequence of positive integers $s \equiv 1 \bmod |G|$ such that any hyperelementary subgroup of Γ_s which projects onto G projects isomorphically to G. (If G is odd, then possibility 2 can be omitted.)*

REMARK. In 2.3, Γ_s denotes $\Gamma/_s\Gamma$; $\pi \rtimes T$ denotes the semidirect product where T is the quotient group, and $*$ is the symbol for amalgamated free product.

Let M be a flat Riemannian manifold with $\pi_1 M = \Gamma$.

2.4. THEOREM (Epstein and Shub [**1968**]). *For any positive integer $s \equiv 1 \bmod |G|$, there exists an expanding endomorphism $f: M \to M$ such that*

1. $|df(X)| = s|X|$ *for all vectors X tangent to M, and*
2. $f_\#: \Gamma \to \Gamma$ *induces multiplication by s on A and the identity map on G.*

REMARK. In 2.4, $f_\#$ denotes the endomorphism of $\pi_1 M$ induced by f.

SKETCH OF PROOF. First construct a homomorphism $\theta: \Gamma \to \Gamma$ which satisfies property 2 in which $f_\#$ is replaced by θ. This is done by a group cohomology argument using the fact that $|G|$ annihilates $H^2(G; A)$. Identify Γ with the group of deck transformations of the universal cover $\mathbb{R}^n$ of M, where $n = \dim M$. Then use 1.23 to produce an affine motion F of $\mathbb{R}^n$ with $F\gamma F^{-1} = \theta(\gamma)$ for all $\gamma \in \Gamma$. Note that $F(x)$ must be $sx + v$ for some $v \in \mathbb{R}^n$ (and all $x \in \mathbb{R}^n$). The posited expanding endomorphism f is induced from F. $\square$

Given a monomorphism $f: \Pi \to \Gamma$ with the index $[\Gamma : f(\Pi)]$ finite, there are transfer maps $f^*: \operatorname{Wh}\Gamma \to \operatorname{Wh}\Pi$ and $f^*: \widetilde{K}_0\mathbb{Z}\Gamma \to \widetilde{K}_0\mathbb{Z}\Pi$ defined algebraically by forgetting information in the $\widetilde{K}_0$ case; i.e., a $\mathbb{Z}\Gamma$-module is also a $\mathbb{Z}\Pi$-module. On the geometric level, let W be a h-cobordism and $\widetilde{W}$ be its covering space corresponding to the subgroup $f(\Pi)$, then $\tau(\widetilde{W}) = f^*\tau(W)$. If $f: M \to M$ satisfies 2.4 relative to a positive integer s, then $f_\#: \pi_1 M \to \pi_1 M$ is called s-expansive.

2.5. GEOMETRIC VANISHING LEMMA. *To each $b \in \operatorname{Wh}\Gamma$ there corresponds a positive integer N_b such that if $\theta: \Gamma \to \Gamma$ is s-expansive (where $s \geq N_b$), then $\theta^*(b) = 0$.*

REMARK. This is a consequence of a basic control theorem due to Steve Ferry. To state it, we then need the following definition.

2.6. DEFINITION. An h-cobordism W, from base N^- to top N^+, is ϵ-controlled provided there exist deformation retractions $r_t^+: W \times I \to N^+$ and $r_t^-: W \times I \to N^-$ (I denotes $[0, 1]$ as usual) such that each of the following paths α_x^+, α_x^- in N^- has diameter less than ϵ:

$$\alpha_x^+(t) = r_1^-(r_t^+(x)) \text{ and } \alpha_x^-(t) = r_1^-(r_t^-(x)).$$

The paths $\{\alpha_x^+, \alpha_x^- | x \in W\}$ are called the tracks of the deformation retractions.

REMARK. Figure 1 illustrates 2.6. The path in W is $t \mapsto r_t^-(x)$ and the path (loop) in N^- is α_x^-. Although N^- is a subspace of W, we place it below W to clarify the illustration.

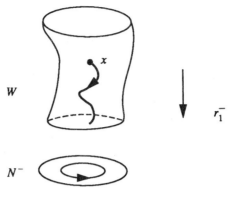

$$W$$

$$N^-$$

FIGURE 1

2.7. THEOREM (Ferry [**1977**]). *Given a closed Riemannian manifold N (with $\dim N > 4$) there exists a positive real number ϵ such that any ϵ-controlled h-cobordism W with base N has $\tau(W, N) = 0$.*

PROOF OF 2.5. Let ϵ be Ferry's number for the flat Riemannian manifold N (with $\pi_1 N = \Gamma$) and W be an h-cobordism on N with $\tau(W) = b$. Let r_t^+ and r_t^- be smooth deformation retractions of W onto N^+ and $N^- = N$, respectively. And let l be an upper bound for the lengths of the tracks of r_t^+ and r_t^-. Choose N_b to be bigger than ℓ/ϵ and we are done. This is because $\theta^* \tau(W) = \tau(\widetilde{W})$ where the base of $\widetilde{W}$ is N. And $\widetilde{W}$ can be equipped with deformation retractions $\tilde{r}_t^+$ and $\tilde{r}_t^-$ whose tracks are lifts of the tracks of r_t^+ and r_t^- relative to the s-expanding endomorphism $f: N \to N$ where $\theta = f_\#$. $\square$

There is also a $\tilde{K}_0$-version of 2.5.

2.8. DEFINITION. Let F be a finite quotient group of Γ and $\varphi: \Gamma \to F$ be the canonical surjection. If E is a subgroup of F, then $\sigma_E: \varphi^{-1}(E) \to \Gamma$ and $\tau_E: E \to F$ denote the inclusion maps.

2.9. ALGEBRAIC VANISHING LEMMA. *Let $\varphi: \Gamma \to F$ be a finite quotient of Γ, then an element b in $\mathrm{Wh}\,\Gamma$ is 0 provided $\sigma_E^*(b) = 0$ for each hyperelementary subgroup E of F.*

PROOF. For each finite subgroup E of F, Swan defined a "functorial" action of a ring $G_0(\mathbb{Z}E)$ on $\tilde{K}_0(\mathbb{Z}\varphi^{-1}(E))]$, cf. Swan [**1970**]. The technical term is that $\tilde{K}_0[\mathbb{Z}\varphi^{-1}(E)]$ is a Frobenius module over the Frobenius functor $G_0(\mathbb{Z}E)$. In particular,

$$\sigma_*(r\sigma^*(b)) = \tau_*(r)b$$

for each pair of elements $r \in G_0(\mathbb{Z}E)$ and $b \in \tilde{K}_0\mathbb{Z}\Gamma$, where $\sigma = \sigma_E$ and $\tau = \tau_E$. But Swan also shows that the multiplicative identity 1 in $G_0(\mathbb{Z}F)$ is a sum

$$1 = \sum \tau_{E^*}(r_E)$$

where $r_E \in G_0(\mathbb{Z}E)$ and the sum ranges over the hyperelementary subgroups of F. Concatenating these two formulas leads directly to 2.9. $\square$

PROOF OF 2.2. We proceed by (double) induction on both rank Γ and $|G|$. First, suppose that case 1 of 2.3 holds; i.e., $\Gamma = \Pi \rtimes T$. In this situation, we can apply the generalization of 1.29 proven in Farrell and Hsiang [1970] to calculate $\mathrm{Wh}\,\Gamma$. Since $\mathbb{Z}\Pi$ is a regular ring (i.e., it is left Noetherian and has finite cohomological dimension) there is an exact sequence

$$\mathrm{Wh}\,\Pi \to \mathrm{Wh}\,\Gamma \to \tilde{K}_0\mathbb{Z}\Pi.$$

But, rank $\Pi = \mathrm{rank}\,\Gamma - 1$. Consequently, both $\mathrm{Wh}\,\Pi$ and $\tilde{K}_0\mathbb{Z}\Pi$ vanish by our inductive assumption. Hence $\mathrm{Wh}\,\Gamma$ vanishes.

Next, suppose case 2 of 2.3 holds; i.e., $\Gamma = B \underset{C}{*} D$. By Waldhausen [1973], there is an exact sequence

$$\mathrm{Wh}\,B \oplus \mathrm{Wh}\,D \to \mathrm{Wh}\,\Gamma \to \tilde{K}_0\mathbb{Z}C$$

because $\mathbb{Z}C$ is a regular ring. Since the ranks of B, C and D are each one less than rank Γ, we again see that the exact sequence combined with our inductive hypotheses yields that $\mathrm{Wh}\,\Gamma$ vanishes.

Finally suppose case 3 of 2.3 holds and let $b \in \mathrm{Wh}\,\Gamma$. Pick an integer s bigger than the number N_b, posited in 2.5, and satisfying the following two conditions. First, $s \equiv 1 \bmod |G|$; second, any hyperelementary subgroup of Γ_s which projects onto G, projects isomorphically to G. Let $h: M \to M$ be a map satisfying 2.4 with respect to s. Now apply 2.9 to the finite quotient $\varphi: \Gamma \to \Gamma_s$. Let E be a hyperelementary subgroup of Γ_s; we need to show that $\sigma_E^*(b) = 0$.

If E does not project onto G, then the holonomy group of $\varphi^{-1}(E)$ has fewer elements than G. In this case, our inductive hypothesis shows $\mathrm{Wh}(\varphi^{-1}(E))$ vanishes; consequently, $\sigma_E^*(b) = 0$. Otherwise, E projects isomorphically to G. All such subgroups are conjugate, since $H^1(G; A/sA) = 0$, and $\varphi f_\#(\Gamma)$ is one of them. Hence it suffices to show $\sigma_E^*(b) = 0$ where $E = \varphi f_\#(\Gamma)$. In this case, $\varphi^{-1}(E) = f_\#(\Gamma)$ and the desired conclusion is a consequence of 2.5, where $\theta = f_\#$.

This completes the proof of 2.2 except for the comment that one must argue the vanishing of $\mathrm{Wh}\,\Gamma$ and $\tilde{K}_0\mathbb{Z}\Gamma$ in an intertwined manner. But this is only a minor difficulty. (See Farrell and Hsiang [1978] for more details.) $\square$

2.10. DEFINITION. A closed manifold is an *infrasolvmanifold* provided it is a double coset space $\Gamma \backslash G / K$ where G is a Lie group which is both virtually connected and virtually solvable, K is a maximal compact subgroup of G and Γ is a torsion-free, cocompact, discrete subgroup of G. When the word solvable is replaced by nilpotent or abelian, we have the definition a closed *infranilmanifold* or closed *infrabelian* manifold; respectively.

2.11. DEFINITION. A group Γ is a *poly-$\mathbb{Z}$ group* provided it has a finite normal series

$$1 = \Gamma_0 \subset \Gamma_1 \subset \cdots \subset \Gamma_n = \Gamma$$

such that Γ_i/Γ_{i-1} is infinite cyclic for $i = 1, 2, \ldots, n$.

The following result can be deduced from Bieberbach [1910], Zassenhaus [1948], Malcev [1951], Mostow [1957] and a construction due to F. E. A. Johnson; cf. Auslander and Johnson [1976].

2.12. THEOREM. *The class of fundamental groups of closed infrabelian manifolds, infranilmanifolds, infrasolvmanifolds, is identical with the class of finitely generated virtually abelian, virtually nilpotent or virtually poly-$\mathbb{Z}$ groups, respectively. Furthermore, every closed infrasolvmanifold is aspherical.*

The technique of proof of 2.2 was extended in Farrell and Hsiang [1981] to yield the following result.

2.13. THEOREM. *If Γ is a virtually poly-$\mathbb{Z}$ group, then* $\mathrm{Wh}\,\Gamma = 0$.

REMARK. The proof of 2.13 requires a more complicated inductive argument than that of 2.2. It uses Auslander and Johnson [1976] to construct singular fibrations of a compact infrasolvmanifold with $\tau_1 M = \Gamma$ over a flat Riemannian orbifold (of positive dimension). Then the arguments proving 2.2 are modified to apply to this setup. In particular, a fibered version of 2.7 is crucial. We proceed to formulate this fibered controlled h-cobordism theorem.

Let $p: N \to B$ be a smooth fiber bundle with fiber F where F, N and B are closed smooth manifolds and B is equipped with a Riemannian metric.

2.14. DEFINITION. A h-cobordism W with base N is ϵ-*controlled relative* to p, provided there exist deformation retractions r_t^+ and r_t^-, as in 2.6, such that the image of the composite $p \circ \alpha$ has diameter less than ϵ in B for each trace α of r_t^+ and r_t^-.

2.15. THEOREM (Quinn [1979]). *Given a closed Riemannian manifold B, there exists a positive number ϵ such that the following is true. Let $p: N \to B$ be a smooth fiber bundle whose fiber is a closed manifold F and where $\dim N > 4$. If $\mathrm{Wh}(\pi_1 F \times A) = 0$ for any free abelian group A, then any h-cobordism W with base N which is ϵ-controlled relative to p has $\tau(W) = 0$.*

The ideas used in calculating $\mathrm{Wh}\,\pi_1 M$ when M is flat Riemannian also yield $|\mathscr{S}(M)| = 1$, when the holonomy group of $\pi_1 M$ has odd order and $\dim M \neq 3, 4$. This was also proven in Farrell and Hsiang [1978]. The same idea eventually led, in Farrell and Hsiang [1983], to proving $|\mathscr{S}(M)| = 1$ without any constraints on the holonomy group of $\pi_1 M$ but still assuming $\dim M \neq 3, 4$. (Recent work of Freedman and Quinn [to appear] allows the restriction $\dim M \neq 4$ to be removed.) The removal of the holonomy constraint also requires using a structure set analogue of 2.15; cf. Farrell and Hsiang [1983], in finessing a certain codimension-one splitting problem which

Cappell [1976] showed was a priori obstructed. Farrell and Jones [1988] prove a structure set (fibered) foliated control theorem where B is foliated with one-dimensional leaves. Then building in a yet more complicated fashion on the arguments used to prove 2.2, they obtain the following result.

2.16. THEOREM (Farrell and Jones [1988]). *Let M be a closed aspherical manifold with $\pi_1 M$ virtually poly-$\mathbb{Z}$. Then $|\mathscr{S}(M)| = 1$ provided* $\dim M \neq 3$.

REMARK. When $\pi_1 M$ is a poly-$\mathbb{Z}$ group "on the nose", 2.16 was proven by Wall [1970]. The results of Freedman and Quinn [to appear] are needed to handle the case when $\dim M = 4$.

2.17. DEFINITION. A *complete flat affine manifold* is the orbit space of affine space $\mathbb{R}^n$ under a properly discontinuous action of a discrete torsion free subgroup of the group G consisting of all affine motions of $\mathbb{R}^n$. Recall G is the semidirect product of $\mathrm{GL}_n(\mathbb{R})$ with the normal subgroup consisting of all translation of $\mathbb{R}^n$.

Milnor [1977] made the following conjecture.

2.18. CONJECTURE. The class of fundamental groups of compact, complete affine flat manifolds is the same as the class of torsion-free, virtually poly-$\mathbb{Z}$ groups.

REMARK. Milnor [1977] showed that any torsion-free virtually poly-$\mathbb{Z}$-group is the fundamental group of some complete flat affine manifold M. But it is unknown, in general, whether M can also be compact except that Boyom [1984] showed this is true when Γ is virtually nilpotent. Some doubts about 2.18 are the consequence of an example, constructed by Margulis [1983], of a complete flat affine manifold N whose fundamental group is not virtually poly-$\mathbb{Z}$. However, N is not compact.

2.19. DEFINITION. A smooth manifold M is *almost flat* provided it supports a sequence g_n of Riemannian metrics such that the sectional curvatures of (M, g_n) converge uniformly to 0 while the diameters of (M, g_n) are uniformly bounded from above.

Ruh [1982], improving on a result of Gromov [1978], obtained the following characterization of almost flat manifolds.

2.20. THEOREM. *The class of closed smooth manifolds which support an almost flat structure is identical with the class of closed infranilmanifolds.*

The above results yield the following *topological* characterizations of closed infrabelian, infranil, and infrasolvmanifolds, respectively.

2.21. COROLLARY. *Let M be a closed topological manifold with* $\dim M \neq 3$, *then the following is true.*

(1) *M has an infrabelian-structure if and only if M has a flat Riemannian structure if and only if M is aspherical and $\pi_1 M$ is virtually abelian.*

(2) *M has an infranil-structure if and only if M has an almost flat structure if and only if M is aspherical and $\pi_1 M$ is virtually nilpotent.*

(3) *M has an infrasolv-structure if and only if M is aspherical and $\pi_1 M$ is virtually solvable. If furthermore 2.18 is true, then M has a complete flat affine structure if and only if M is aspherical and $\pi_1 M$ is virtually solvable.*

REMARKS. Parts (1) and (2) of 2.21 are proved in Farrell and Hsiang [**1983**] while (3) can be deduced, in the same way, from 2.12 and 2.16. We emphasize the characterizations in 2.21 are topological and *not* smooth. For example, $T^n \# \Sigma^n$ (when $n > 4$) supports a flat Riemannian structure only when Σ^n is diffeomorphic to S^n. (Here T^n denotes the n-dimensional torus and Σ^n is a smooth manifold homeomorphic to S^n.)

We close this chapter with some remarks about the Nil functor. It is necessary, because of 1.29, that the following conjecture be true for either 1.21 or 1.22 to be true.

2.22. NIL CONJECTURE. *If Γ is any torsion-free group, then $\operatorname{Nil} \mathbb{Z}\Gamma = 0$.*

2.23. THEOREM. *Let R denote a ring with 1, then the following are true.*
(1) *If $\operatorname{Nil} R \neq 0$, then $\operatorname{Nil} R$ is not finitely generated.*
(2) *If R is a regular ring, then $\operatorname{Nil} R = 0$.*
(3) *Let Γ be a finitely generated discrete subgroup of a virtually connected Lie group. Then $\mathbb{Z}\Gamma$ is regular if and only if Γ is a torsion-free virtually poly-$\mathbb{Z}$ group.*
(4) *Let p be any prime and $\Gamma = T \times C_p \times C_p$, then $\operatorname{Nil}(\mathbb{Z}\Gamma)$ is not finitely generated.*

REMARKS. Part (1) of 2.23 is proven in Farrell [**1977**] and (2) in Bass, Heller and Swan [**1964**]. Part (3) is proven in Milnor [**1977**]. He deduces it from Tits [**1972**] and Mostow [**1957**]. Part (4) is proven in Bass and Murthy [**1967**].

3. The Algebraic K-theory of Hyperbolic Manifolds

Hyperbolic manifolds have an alternate synthetic geometry description similar to the definition of complete affine flat manifolds given in 2.17. Namely, a n-dimensional hyperbolic manifold is the orbit space of hyperbolic n-space $\mathbb{H}^n$ under the action of a torsion-free discrete subgroup of the Lie group $\mathrm{Iso}\,\mathbb{H}^n$ consisting of all isometries of $\mathbb{H}^n$. There are several models for $\mathbb{H}^n$; we briefly describe two of them.

In the first model, $\mathbb{H}^n$ is identified with following subset of $\mathbb{R}^{n+1}$

$$\{(x_1, x_2, \ldots, x_{n+1}) | x_1^2 + x_2^2 + \cdots + x_n^2 - x_{n+1}^2 = -1, \text{ and } x_{n+1} > 0\}.$$

The Riemannian metric on $\mathbb{H}^n$ is induced from the flat pseudo-Riemannian metric on $\mathbb{R}^{n+1}$ naturally determined by the $(n+1) \times (n+1)$ symmetric matrix D where

$$D_{ij} = \begin{cases} 0 & \text{if } i \neq j \\ 1 & \text{if } i = j \text{ and } i \leq n \\ -1 & \text{if } i = j = n+1 \end{cases}$$

It is clear, using this model for $\mathbb{H}^n$, that $\mathrm{Iso}\,\mathbb{H}^n$ can be identified with the group $\mathrm{O}^+(n, 1, \mathbb{R})$; i.e., with all $(n+1) \times (n+1)$ matrices A satisfying the two conditions

$$ADA^t = D \text{ and}$$

the bottom entry of Ae_{n+1} is positive

where e_{n+1} denotes the column vector of length $n+1$ whose bottom entry is 1.

In the Poincaré model, $\mathbb{H}^n$ is identified with the interior of the unit ball $\mathbb{B}^n$ centered at 0. Let $x_1, \ldots, x_n$ be the usual coordinate functions on $\mathbb{R}^n$ and let $X_1 = \partial/\partial x_i$ for $i = 1, 2, \ldots, n$. The Riemannian metric $\langle \, , \, \rangle$ on this model of $\mathbb{H}^n$ is described by

$$\langle X_i, X_j \rangle = 4\delta_{ij}/A^2 \text{ where}$$

$$\delta_{ij} = \begin{cases} 0 & \text{if } i \neq j \\ 1 & \text{if } i = j \end{cases}$$

and

19

$$A = 1 - \sum_{i=1}^{n} (x_i)^2$$

Using this model, it can be seen that $\mathbb{B}^n$ is a compactification of $\mathbb{H}^n$ and Iso $\mathbb{H}^n$ extends to an action on $\mathbb{B}^n$. (See Mostow [1967] for more details.)

We will return to these models later in this chapter during the proof of the following result.

3.1. THEOREM (Farrell and Jones [1986]). *If M is a hyperbolic manifold, then* Wh $\pi_1 M = 0$. (*More generally*, Wh$(\pi_1 M \times A) = 0$ *for any free abelian group A. Hence also both* $\tilde{K}_0(\mathbb{Z}\pi_1 M) = 0$ *and* Nil$(\mathbb{Z}\pi_1 M) = 0$.)

REMARKS. The class of fundamental group of hyperbolic manifold is identical to the class of discrete, torsion-free subgroups of $O(n, 1)$ where n varies over all positive integers. (Note that $O(n, 1)$ is the direct product of $O^+(n, 1)$ and the cyclic group of order two generated by $-I$, where I denotes the identity matrix.) Thus 3.1 states that Wh $\Gamma = 0$ for any discrete, torsion-free subgroup of $O(n, 1)$. The following results were known prior to the proof of 3.1. Farrell and Hsiang [1970], and independently Waldhausen [1973], showed Wh $\pi_1 M = 0$ when M is a two-dimensional manifold. Waldhausen [1973] also showed this when M is a sufficiently large three-dimensional manifold. Nicas and Stark [1984], [1987] showed this for a certain (up to then) intractible class of nonsufficiently large hyperbolic three-manifolds.

We now proceed to sketch the proof of 3.1 for the special case when M is compact. We start by explaining a key ingredient; i.e., the *asymptotic transfer* of a path $\alpha: I \to M$ to a path $v\alpha$ in SM, where

$$p: SM \to M$$

denotes the sphere bundle consisting of all unit length vectors tangent to M, and $v \in p^{-1}(\alpha(0))$. (We define it even when M is not compact.) The asymptotic transfer sits on top of α in the sense that composite $p \circ (v\alpha)$ is α. We now describe its construction. If $v \in SM$, let γ_v denote the geodesic such that

$$\dot{\gamma}_v(0) = v.$$

The distance induced by the Riemannian metric on M is denoted by $d(,)$. Two geodesics γ and σ in $\mathbb{H}^n$ are asymptotes if $d(\gamma(t), \sigma(t)) \leq C$ for some constant $C > 0$ and all $t \geq 0$, i.e., they stay a finite distance apart in positive time. Vectors $v, w \in S\mathbb{H}^n$ are asymptotes if γ_v and γ_w are asymptotes. Using the Poincaré model for $\mathbb{H}^n$, we can concretely visualize these objects as follows. The geodesic γ_v is contained in the circle (or straight line in the degenerate case) which is tangent to v and meets S^{n-1} perpendicularly in points γ_v^+ and γ_v^-. The positive ray $\{\gamma_v(t)|t \geq 0\}$ accumulates to the point γ_v^+ and the vectors v and w are asymptotes provided $\gamma_v^+ = \gamma_w^+$.

We first define $v\alpha$ in the special case when $M = \mathbb{H}^n$. The general case then follows using the fact that M is the orbit space of $\mathbb{H}^n$ under the action of a group Γ of isometries of $\mathbb{H}^n$ and observing that the construction, in the special case, is Γ-equivariant. For each $v \in S\mathbb{H}^n$, we denote by $v(x)$ the unique vector at x such that γ_v and $\gamma_{v(x)}$ are asymptotes. Then $v\alpha$ is defined by the formula

$$v\alpha(t) = v(\alpha(t))$$

for $t \in [0, 1]$. Figure 2 illustrates this construction.

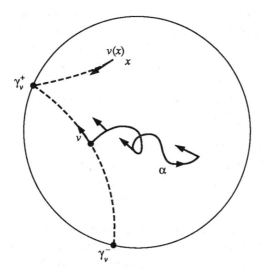

FIGURE 2

The following are some properties of the asymptotic transfer.

(1) If α is a null homotopic loop, then so is $v\alpha$.

(2) If α is a constant loop, so is $v\alpha$.

(3) If α is a smooth curve, $v\alpha$ is one also. Furthermore, $\|v\dot\alpha(t)\| \leq \sqrt{2}\|\dot\alpha(t)\|$ for each $t \in [0, 1]$.

REMARK. There is a natural Riemannian metric on SM induced from that on M. The inequality in property (3) is stated in terms of this pair of metrics. (See Farrell and Jones [**1986**, p. 547] for more details.)

There is a natural flow g^t, $t \in \mathbb{R}$, defined on SM by

$$g^t(v) = \gamma_v(t)$$

called the *geodesic flow*. The key property of the asymptotic transfer is that the geodesic flow shrinks $v\alpha$ in every direction except the flow line direction; i.e., it deforms it arbitrarily close to a flow line (as $t \to +\infty$) while keeping it bounded above in length. (A glance at Figure 2 should make this property plausible. One can verify it by using the fact that g^t is an Anosov flow and $v\alpha$ is contained in a weakly stable leaf of this flow.) We now quantitatively formulate this property.

3.2. DEFINITION. A path γ in SM is (β, ϵ)-*controlled* provided there exists a second path φ in SM satisfying the following properties.

(1) The image of φ is contained in an arc of length β inside a flow line of the geodesic flow.

(2) $d(\gamma(t), \varphi(t)) < \epsilon$ for all $t \in [0, 1]$.

REMARK. We use $d(\ ,\ )$ to denote the distance function on a Riemannian manifold. An h-cobordism (W, SM) is (β, ϵ)-*controlled* provided it can be equipped with deformation retractions whose tracks are all (β, ϵ)-controlled.

3.3. KEY PROPERTY OF $v\alpha$. Given positive real numbers β and ϵ, there exists a positive number s_0 such that the following is true. Let α be any smooth path in M whose arc length is less than β and v any unit length vector tangent to M at $\alpha(0)$. Then the composite $g^s \circ (v\alpha)$ is $(\sqrt{2}\beta, \epsilon)$-controlled, provided $s \geq s_0$.

This fact is useful because of the following generalization of 2.7, proven in Farrell and Jones [1986].

3.4. FOLIATED CONTROL THEOREM (special case). *Given a closed hyperbolic manifold M with* $\dim M > 2$ *and a positive real number β, there exists a number $\epsilon > 0$ such that the following is true. Every (β, ϵ)-controlled h-cobordism with base SM has zero Whitehead torsion; i.e., is a product.*

REMARK. The proof 3.4 follows the pattern given in Ferry [1977] for 2.7, once a *long-thin cell structure* $\mathscr{C}$ is constructed for SM. Roughly speaking this means that, for each cell σ of $\mathscr{C}$, there is a smooth chart $f: \mathbb{R} \times \mathbb{R}^{2n-2} \to SM$ (where $n = \dim M$) with $f(\mathbb{R} \times x)$ contained in a flow line of g^t, for each $x \in \mathbb{R}^{2n-2}$. There is also a pair of cells σ_1 and σ_2 in $\mathbb{R}$ and $\mathbb{R}^{2n-2}$, respectively, with $\sigma = f(\sigma_1 \times \sigma_2)$ such that $f(\sigma_1 \times x)$ is "long" for each point $x \in \sigma_2$, provided $\dim \sigma_1 = 1$. And $f(t \times \sigma_2)$ is "skinny" for each $t \in \sigma_1$. Furthermore, σ_2 is triangulated by a piecewise smooth triangulation of $\mathbb{R}^{2n-2}$. (See Farrell and Jones [1986], pages 560–564 for more details.)

We now relate these facts to 3.1. Let $x \in \mathrm{Wh}\,\Gamma$; (W, M) be a smooth h-cobordism with $\tau(W) = x$, and h_t^+, h_t^- be smooth deformation retractions of W onto its top and base, respectively. Let β be an upper bound for the arc lengths of the tracks of h_t^+ and h_t^-. (It is quite unlikely that β is smaller than the number ϵ posited in 2.7 in M. Hence we cannot a priori conclude that $x = 0$.) Let $\mathscr{W}$ be the total space of the pullback of $p: SM \to M$ via h_1^-. Then $(\mathscr{W}, SM)$ is an h-cobordism which can be (easily) equipped with smooth deformation retractions r_t^+ and r_t^- onto its top and base, respectively, such that the tracks of r_t^+ and r_t^- are the asymptotic transfers of the tracks of h_t^+ and h_t^-. Furthermore, given a selfdiffeomorphism $f\colon SM \to SM$, we can modify r_t^+ and r_t^- to deformation retractions whose tracks are the composites of f with the tracks of r_t^+ and r_t^-. Let ϵ be the number determined in 3.4 for M and $\sqrt{2}\beta$; s_0 be the number posited in 3.3 relative to β and ϵ, and g^s be the diffeomorphism f of the previous sentence, where s is a number greater than s_0. Thus we see $(\mathscr{W}, SM)$ is $(\sqrt{2}\beta, \epsilon)$-controlled. Hence 3.4 implies that $\tau(\mathscr{W}) = 0$. The next result relates $\tau(W)$ to $\tau(\mathscr{W})$.

3.5. THEOREM (Anderson [1972]). *Let (W, M) and $(\mathscr{W}, \mathscr{M})$ be h-cobordisms such that $p\colon \mathscr{W} \to W$ is a smooth fiber bundle with $p^{-1}(M) = \mathscr{M}$ and $\dim M > 4$. Assume that $\pi_1 W$ acts trivially on the integral homology groups of the fiber F of p, then*

$$p_*(\tau(\mathscr{W})) = \chi(F)\tau(W)$$

where $\chi(F)$ denotes the Euler characteristic of F and $p_*\colon \mathrm{Wh}\,\pi_1\mathscr{M} \to \mathrm{Wh}\,\pi_1 M$ is the homomorphism induced by p.

By applying 3.5 to the h-cobordism $\mathscr{W}$ constructed above, we see that

$$\chi(S^{n-1})x = 0$$

where $n = \dim M$, provided M is orientable. Consequently, $2x = 0$ for every element $x \in \mathrm{Wh}\,\pi_1 M$ when M is orientable and odd dimensional. (If M is even dimensional and orientable, we apparently get no information about $\mathrm{Wh}\,\pi_1 M$ from the discussion so far.)

This argument fails to prove 3.1 because $\chi(S^n) \neq 1$. To get around this difficulty, we look for a subbundle of SM with an asymptotic transfer and whose fiber F has $\chi(F) = 1$. Identify SM with the balanced product $\mathbb{H}^n \times_\Gamma S^{n-1}$ where Γ is the subgroup of $\mathrm{Iso}\,\mathbb{H}^n$ such that $M = \mathbb{H}^n/\Gamma$, where we use the Poincaré model for $\mathbb{H}^n$. This identification, when $M = \mathbb{H}^n$, is given by

$$v \mapsto (\gamma_v(0), \gamma_v^+)$$

where $v \in S\mathbb{H}^n$, and the geodesic flow becomes

$$g^t(\gamma_v(0), \gamma_v^+) = (\gamma_v(t), \gamma_v^+).$$

Since this function is $\mathrm{Iso}\,\mathbb{H}^n$-equivariant, it induces an identification of SM with $\mathbb{H}^n \times_\Gamma S^{n-1}$. The subbundle should be $\mathbb{H}^n \times_\Gamma F$ where F is a Γ-invariant closed subspace of S^{n-1}. Note that $\mathbb{H}^n \times_\Gamma F$ is a subspace of SM which is left invariant under the geodesic flow and contains the asymptotic transfer $v\alpha$ of any curve α in M, provided $\gamma_v^+ \in F$. Unfortunately each orbit of Γ is dense in S^{n-1} when M is compact. We are thus forced to consider a certain noncompact hyperbolic manifold N called the *enlargement* of M. It contains M as a totally geodesic codimension-1 subspace and is diffeomorphic to $M \times \mathbb{R}$. We give two descriptions of N. The first is in terms of the Poincaré model for $\mathbb{H}^{n+1}$, where we identify $\mathbb{H}^n$ with the totally geodesic subspace given by $\mathbb{H}^n = \mathbb{R}^n \cap \mathbb{H}^{n+1}$. Embed $\mathrm{Iso}\,\mathbb{H}^n = O^+(n, 1)$ into $\mathrm{Iso}\,\mathbb{H}^{n+1} = O^+(n+1, 1)$ by sending each matrix A in $O^+(n, 1)$ to the corresponding blocked matrix

$$\begin{pmatrix} 1 & 0 & \cdots & 0 \\ 0 & & & \\ \vdots & & A & \\ 0 & & & \end{pmatrix}$$

in $O^+(n+1, 1)$. Thus Γ is identified with a discrete, but no longer cocompact, subgroup of $\mathrm{Iso}\,\mathbb{H}^{n+1}$, then $N = \mathbb{H}^{n+1}/\Gamma$ is the enlargement of M. The closed northern hemisphere S^+ of S^n is clearly Γ-invariant and $\chi(S^+) = 1$ since S^+ is homeomorphic to $\mathbb{B}^n$. This is the desired fiber F; but relative to N not M. The smooth manifold with boundary $\mathbb{H}^{n+1} \times_\Gamma S^+$ is denoted S^+M. It is the total space of the fiber bundle

$$p: S^+M \to N$$

with fiber S^+, where p is induced by the projection onto the first factor of $\mathbb{H}^{n+1} \times S^+$.

The second description of N is $M \times (-1, 1)$ equipped with the Riemannian metric $B(\ ,\ )$ defined by

$$B_{(x,t)}(X, Y) = \frac{1}{1-t^2} b_x(X_0, Y_0) + \frac{rs}{(1-t^2)^2}$$

where $X, Y \in T_{(x,t)}M \times (-1, 1)$ and

$$dp_1(X) = X_0, \qquad dp_1(Y) = Y_0,$$
$$dp_2(X) = r\frac{\partial}{\partial t}, \qquad dp_2(Y) = s\frac{\partial}{\partial t} \text{ with}$$
$$p_1: M \times (-1, 1) \to M, \qquad p_2: M \times (-1, 1) \to (-1, 1)$$

denoting the canonical projections onto the first and second factors, respectively. The original manifold M is identified with the totally geodesic submanifold $M \times 0$ of its enlargement $M \times (-1, 1)$. Figure 3 illustrates the second description of the enlargement N of M.

There is a more general notion of h-cobordism which allows the base to be a noncompact manifold with boundary.

3.6. DEFINITION. A manifold W is a (compactly supported) *h-cobordism* with top N^+ and base N provided the following is true.

1. Both N^+ and N are disjoint codimension-0 submanifolds of ∂W.

2. There exist deformation retractions r_t^+ and r_t^- of W onto N^+ and N, respectively.

3. Let $\mathcal{N}$ be ∂W with the union of the interiors of N^+ and N deleted. Then r_t^+ and r_t^- restrict to give deformation retractions of $\mathcal{N}$ onto ∂N^+ and ∂N, respectively.

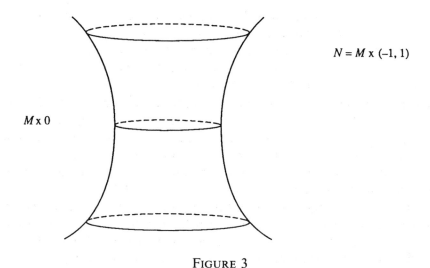

$N = M \times (-1, 1)$

$M \times 0$

FIGURE 3

4. There is a compact subset K of N with $(r_1^-)^{-1}(N - K) = (N - K) \times [0, 1]$ and $N - K$ is identified with $(N - K) \times 0$.

5. It is required that $r_t^-(y, s) = (y, (1-t)s)$ and $r_t^+(y, s) = (y, s+t(1-s))$ for all s, $t \in [0, 1]$ and $y \in N - K$.

REMARKS. Let N' be any compact codimension-0 submanifold of N containing the set K, of property (4), in its interior and let $W' = (r_1^-)^{-1}(N')$. The torsion $\tau(W)$ of W is defined to be $\sigma_*(x)$ where $\sigma: N' \to N$ is the inclusion map and x is the Whitehead torsion of $r_1^-: W' \to N'$. (The germ of the product structure at ∞ is part of the h-cobordism W; hence $\tau(W)$ is independent of the choice of N'.) Every element in $\mathrm{Wh}\,\pi_1 N$ is the torsion of some h-cobordism with base N. An h-cobordism with base S^+M is said to be (β, ϵ)-controlled provided the deformation retractions r_t^+ and r_t^- can be chosen so that their tracks are (β, ϵ)-controlled paths in SN. The total space $\mathscr{W}$ of the pullback over W of the bundle $p: S^+M \to N$ via r_1^- is an h-cobordism with $p_*(\tau(\mathscr{W})) = \tau(W)$. This can be seen directly without using 3.5 (or a generalization of 3.5).

Here is a relevant result related to 3.4.

3.7. THEOREM (Farrell and Jones). *Given a closed hyperbolic manifold M and a positive number β, there exists a positive number ϵ such that the following is true. If W is a (β, ϵ)-controlled h-cobordism with base S^+M, then $\tau(W) = 0$.*

REMARK. This result is deduced from the arguments which also yield the main foliated control theorem of Farrell and Jones [1988b] using the fact that the enlargement of M has positive injectivity radius. In fact the injectivity radius of its enlargement equals that of M itself, which can be understood by looking at Figure 3. It also uses that M has only a finite number of closed geodesics of length less than any fixed number and that each deck transformation of its universal cover is semi-simple.

PROOF OF 3.1. Let N denote the enlargement of M and (W, N) be an h-cobordism with $\tau(W) = \sigma_*(x)$, where σ is the map which identifies M to $M \times 0$ in $M \times (-1, 1) = N$. Note that $\sigma_* \colon \mathrm{Wh}\,\pi_1 M \to \mathrm{Wh}\,\pi_1 N$ is an isomorphism.) Let $(\mathscr{W}, S^+ M)$ be the h-cobordism which is the pullback of the bundle $p \colon S^+ M \to M$ via the retraction $r_1^- \colon W \to N$. Then $p_*(\tau(\mathscr{W})) = \sigma_*(x)$, and $\mathscr{W}$ is (β, ϵ)-controlled for some fixed positive number β but ϵ can be arbitrarily small. Thus $\tau(\mathscr{W}) = 0$ because of 3.7. □

REMARK. The original proof of 3.1, in Farrell and Jones [1986], is slightly more complicated since it uses a weaker result than 3.7.

Let $A \subset B$ be compact Hausdorff spaces, then $\mathrm{Homeo}(B, A)$ denotes the space of self-homeomorphisms of B which are fixed on A. This space has the compact open topology. The notation is abbreviated to $\mathrm{Homeo}(B)$ when $A = \varnothing$. Another basic problem, along with 1.2, is the following.

3.8. PROBLEM. Calculate the homotopy groups $\pi_n \mathrm{Homeo}(M)$, when M is a closed manifold.

3.9. DEFINITION. Let M be a compact manifold. The *pseudo-isotopy* space of M is $\mathrm{Homeo}(M \times [0, 1], M \times 0)$ and is denoted by $P(M)$. An element in it is called a pseudoisotopy of M.

3.10. PROBLEM. Calculate $\pi_n P(M)$ when M is a compact manifold.

REMARK. Solving 3.10 would be important step toward solving 3.8. This relation is discussed in Hatcher [1978].

We will come back to it in Chapter 5. There is a stabilized version of $P(M)$. Let I^n denote $[0, 1] \times \cdots \times [0, 1]$, n-factors. Embed $P(M \times I^n)$ into $P(M \times I^{n+1})$ by sending $f \in P(M \times I^n)$ to $f \times \mathrm{id}$ in $P(M \times I^n \times I)$.

3.11. DEFINITION. Let $\mathscr{P}(M)$ denote the $\displaystyle\lim_{n \to +\infty} P(M \times I^n)$. It is called the *stable pseudo-isotopy space* of M.

There is a stable range through which $\pi_n P(M)$ is isomorphic to $\pi_n \mathscr{P}(M)$.

3.12. THEOREM (Igusa [1988]). *The inclusion map of $P(M)$ into $\mathscr{P}(M)$ induces an isomorphism on homotopy groups π_q for all $q \le \frac{\dim M - 7}{3}$, provided* $\dim M > 10$.

REMARK. Igusa [1988] proves the analogue of 3.12 in the smooth category. His proof builds on ideas in Hatcher [1975]. Burghelea and Lashof [1974] showed that stability in the smooth category implies stability in the topological category. Goodwillie showed the stability ranges are identical.

The techniques used to prove 3.1 also yields information about $\mathscr{P}(M)$ when M is a hyperbolic manifold.

3.13. DEFINITION. The topological space $\overline{\mathscr{P}}(S^1)$ is the direct limit as $n \to +\infty$ of the n-factor Cartesian product

$$\mathscr{P}(S^1) \times \cdots \times \mathscr{P}(S^1).$$

That is, $\overline{\mathscr{P}}(S^1)$ is the subspace of the Cartesian product of a countably infinite number of copies of $\mathscr{P}(S^1)$ (with the box topology) consisting of those elements with only a finite number of coordinates different from id $\in$ $\mathscr{P}(S^1)$. (Recall a basic open set in the box topology for the Cartesian product is an arbitrary Cartesian product of open subsets of the factors.)

3.14. THEOREM (Farrell and Jones [1987]). *Let M be a closed hyperbolic manifold. Then $\mathscr{P}(M)$ and $\overline{\mathscr{P}}(S^1)$ are weakly homotopically equivalent; i.e., there exists a continuous map of $\overline{\mathscr{P}}(S^1)$ into $\mathscr{P}(M)$ which induces an isomorphism on π_n for all integers $n \geq 0$.*

REMARK. More complicated versions of 3.14 are true when M is hyperbolic but *not* compact. (See Farrell and Jones [1987] and [1989, Appendix].)

Much is known about $\mathscr{P}(S^1)$; here are two important results.

3.15. THEOREM (Hatcher). *The group $\pi_0\mathscr{P}(S^1)$ is not finitely generated; in fact, it is the direct sum of a countably infinite number of copies of the cyclic group of order 2.*

3.16. THEOREM (Waldhausen [1978]). *For all integers $n \geq 0$, $\pi_n\mathscr{P}(S^1) \otimes \mathbb{Q} = 0$.*

REMARKS. Theorem 3.15 is proven in Hatcher and Wagoner [1973]. See Hatcher [1978, pp. 9–10] for an explicit construction of elements in $\pi_0\mathscr{P}(S^1)$. These elements come from the stable one-stem (i.e., from $\pi_{n+1}(S^n)$ where $n > 2$). It is rather surprising that $\mathscr{P}(S^1)$ is not contractible.

3.17. PROBLEM. Calculate $\pi_n\mathscr{P}(S^1)$ in terms of the stable stems; i.e., in terms of $\pi_{m+i}(S^m)$ where $m \gg i$.

REMARK. Igusa has done this for $\pi_1\mathscr{P}(S^1)$; cf. Hatcher [1978, p. 7].

Concatenating 3.14 with 3.12, 3.15 and 3.16 has the following consequence.

3.18. COROLLARY (Farrell and Jones [1987]). *Let M be a closed hyperbolic manifold. Then*

$$\pi_n\mathscr{P}(M) \otimes \mathbb{Q} = 0 \quad \text{for all } n \geq 0, \text{ and}$$
$$\pi_0\mathscr{P}(M) \text{ is not finitely generated.}$$

In fact, $\pi_0\mathscr{P}(M)$ is a countably infinite direct sum of cyclic groups of order 2. Moreover if $\dim M > 10$ and $n \leq \frac{\dim M - 7}{3}$ then

$$\pi_n P(M) \otimes \mathbb{Q} = 0$$

and $\pi_0 P(M)$ is a countably infinite direct sum of cyclic groups of order 2.

Let R be a ring with identity 1. Quillen [1973] defined abelian group valued functors $K_n R$, $n \geq 2$, generalizing the functors $K_n R$, $n \leq 2$, which had already been constructed. The group $K_n R \otimes \mathbb{Q}$ is determined by the homology of $\mathrm{GL}(R)$ where

$$\mathrm{GL}(R) = \lim_{n \to +\infty} \mathrm{GL}_n R; \text{ in fact}$$

$$K_n R = \pi_n(\mathrm{BGL}(R))^+.$$

The groups $K_n R$, $n < 0$, are defined inductively as the quotient of K_{n+1} $R[t, t^{-1}]$ relative to the subgroup generated by all elements coming from either $K_{n+1} R[t]$ or $K_{n+1} R[t^{-1}]$ under maps induced by the inclusions of $R[t]$ and $R[t^{-1}]$ into $R[t, t^{-1}]$. (This definition is an outgrowth of Bass, Heller and Swan [1964].) Loday [1976] defined, for each integer n, a functor from groups G to abelian groups $\mathrm{Wh}_n G$. It is defined using $K_n \mathbb{Z}G$ and for $n \leq 1$ agrees with earlier definitions, namely

$$\mathrm{Wh}_n(G) = \begin{cases} \mathrm{Wh}\, G & \text{if } n = 1 \\ \tilde{K}_0 \mathbb{Z}G & \text{if } n = 0 \\ K_n \mathbb{Z}G & \text{if } n < 0 \end{cases}$$

3.19. THEOREM (Waldhausen [1978]). *Let M be a compact aspherical manifold, then*

$$\pi_n \mathscr{P}(M) \otimes \mathbb{Q} \simeq \mathrm{Wh}_{n+2}(\pi_1 M) \otimes \mathbb{Q}.$$

3.20. ADDENDUM (Nicas [1985]). *Let M be a compact aspherical manifold and $\nu = [(n+1)/2]!$ where n is an integer bigger than 1. Then $\mathrm{Wh}_n(\pi_1 M) \otimes \mathbb{Z}[\frac{1}{\nu}]$ is a quotient group of $\pi_{n-2}\mathscr{P}(M) \otimes \mathbb{Z}[\frac{1}{\nu}]$.*

REMARK. The symbol $[x]$ denotes the greatest integer in the real number x.

3.21. COROLLARY (Farrell and Jones [1987]). *Let M be a closed hyperbolic manifold and n be an arbitrary integer. Then*

$$\mathrm{Wh}_n \pi_1 M \otimes \mathbb{Z}[\tfrac{1}{\nu}] = 0$$

where $\nu = [(n+1)/2]!$ when $n > 2$, and $\nu = 1$ otherwise.

REMARKS. The following results were known prior to Farrell and Jones [1987]. Waldhausen [1978b] showed $\mathrm{Wh}_n \pi_1 M = 0$, for all n, if M is either an aspherical 2-manifold or a sufficiently large 3-manifold. Nicas and Stark [1984], [1987] and Nicas [1986] proved 3.21 for certain interesting classes of nonsufficiently large 3-manifolds. The derivation of 3.21, from 3.14 and 3.20, uses that the epimorphism of 3.20 is "natural" with respect to the isomorphism of 3.18 and the fact that $\mathrm{Wh}_n(\pi_1 S^1) = 0$ for all integers n from Quillen [1973]. (See Farrell and Jones [1987, pp. 482–483] for details.) Also 3.21 is true for arbitrary hyperbolic manifolds because of Farrell and Jones [1989, Addendum].

There are relevant generalizations of 1.21 and 1.30.

3.22. CONJECTURE. If Γ is a torsion-free group, then $\mathrm{Wh}_n \Gamma = 0$ for all integers n.

3.23 CONJECTURE. For any group G, $K_n \mathbb{Z} G = 0$ provided $n < -1$.

REMARKS. Carter [1980] verified 3.23 when G is a finite group. Using his result and building on the techniques used to prove 3.1, Farrell and Jones [1986b] verified 3.23 for any cocompact discrete subgroup of $O(m, 1)$ where m is arbitrary.

The groups $\mathrm{Wh}_n(G)$ and $K_n(\mathbb{Z} G)$ fit into a long exact sequence, cf. Loday [1976]. This sequence, together with one of the remarks following 3.21, yields the following calculation.

3.24. COROLLARY (Farrell and Jones [1989]). *Let Γ be a discrete, torsion-free subgroup of $O(m, 1)$ for some integer m, then*

$$K_n(\mathbb{Z}\Gamma) \otimes \mathbb{Q} \simeq H_n(\Gamma, \mathbb{Q}) \oplus (\oplus_{i=1}^{\infty} H_{(n-1)-4i}(\Gamma, \mathbb{Q})).$$

Let $P_{Diff}(M)$ denote the space of smooth pseudoisotopies for a compact smooth manifold M. We finish this chapter by stating an application of 3.24 and 3.12 combined with Waldhausen [1978]; which yields quantitative information about $P_{Diff}(M)$.

3.25. THEOREM (Farrell and Jones [1987]). *Let M be a closed hyperbolic manifold with $\dim M > 10$, then*

$$\pi_i P_{Diff}(M) \otimes \mathbb{Q} \simeq \bigoplus_{s=1}^{\infty} H_{(i+1)-4s}(M, \mathbb{Q})$$

provided $i \leq \frac{\dim M - 7}{3}$.

4. Locally Symmetric Spaces of Noncompact Type

4.1. DEFINITION. A simply connected nonpositively curved symmetric space with no (positive dimensional) flat factor is called a *symmetric space of noncompact type*.

Throughout this Chapter, X will denote a symmetric space of noncompact type. Recall that X is diffeomorphic to $\mathbb{R}^m$ where $m = \dim X$.

4.2. DEFINITION. The *rank* of X is the maximal dimension of a flat, totally geodesic subspace in X; rank X is usually denoted by r. A flat, totally geodesic subspace of X with dimension r is called an *r-flat*. Iso(X) denotes the group consisting of all the isometries of X.

Iso(X) is a (virtually connected) semisimple linear Lie group. It acts transitively on both X and the set of all r-flats in X. The isotopy subgroup of a point in X is a maximal compact subgroup K of Iso(X). Thus X can be identified with the homogeneous space Iso(X)/K. This string of reasoning can be reversed. Let $\mathcal{G}$ be a (virtually connected) semisimple linear Lie group and $\mathcal{K}$ be a maximal compact subgroup of $\mathcal{G}$. Let $\mathcal{X}$ denote the homogeneous space $\mathcal{G}/\mathcal{K}$ equipped with a $\mathcal{G}$-invariant Riemannian metric. Then $\mathcal{X}$ is a symmetric space of noncompact type. (See Mostow [1973, §3] for more details.)

4.3. EXAMPLE. The homogeneous space $\mathrm{SL}_n(\mathbb{R})/\mathrm{SO}_n$ (where $n > 1$) is a smooth manifold whose dimension is $2+3+\cdots+n = (n^2+n-2)/2$. Equip it with a $\mathrm{SL}_n(\mathbb{R})$ invariant Riemannian metric (which is unique up to a positive constant multiple). The resulting Riemannian manifold X_n is a symmetric space of noncompact type since $\mathrm{SL}_n(\mathbb{R})$ is semisimple and SO_n is a maximal compact subgroup in it. Provided the right constant is chosen, X_2 is $\mathbb{H}^2$. On the other hand X_n is never hyperbolic, when $n > 2$, since it contains a flat, totally geodesic $n-1$ dimensional subspace. One such subspace is $F = A[I]$ where $[I]$ denotes the coset represented by the identity matrix I and $A[I]$ is the orbit of $[I]$ under the action of the subgroup A consisting of all diagonal matrices in $\mathrm{SL}_n(\mathbb{R})$ with positive diagonal entries. This space F is isometrically equivalent to flat Euclidean $n-1$ dimensional space and is a maximal flat, totally geodesic subspace of X_n. Hence the rank of X_n is $n-1$. The set $\{gF | g \in \mathrm{SL}_n(\mathbb{R})\}$ is identical with the collection of r-flats in

X_n, where $r = n - 1$.

4.4. DEFINITION. A locally symmetric space is of *noncompact type* provided its universal cover is a symmetric space of noncompact type.

REMARKS. Locally symmetric spaces of noncompact type are the orbit space X/Γ where Γ is a torsion-free, discrete subgroup of $\mathrm{Iso}(X)$. They are the double coset space $\Gamma\backslash G/K$ where G is a (virtually connected) semisimple linear Lie group, K is a maximal compact subgroup of G, and Γ is a discrete, torsion-free subgroup of G. The class of fundamental groups of locally symmetric spaces of noncompact type is identical with the class of torsion-free discrete subgroups of (virtually connected) semisimple Lie groups.

4.5. THEOREM (Borel [**1963**]). *Let G be a (virtually connected) semisimple Lie group. Then G contains a torsion-free, discrete, cocompact subgroup. Consequently, any symmetric space X of noncompact type is the universal cover of a locally symmetric space M of noncompact type such that M is a closed manifold and each deck transformation is an isometry of X.*

REMARK. There are counterexamples to 4.5 when G is not semisimple. For example, there are nilpotent Lie groups for which it is false.

4.6. THEOREM (Farrell and Jones [**1987b**]). *Let M be a locally symmetric space of noncompact type which is a closed manifold, then $\mathrm{Wh}(\pi_1 M)\otimes\mathbb{Q} = 0$. (More generally, $\mathrm{Wh}(\pi_1 M \times A)\otimes\mathbb{Q} = 0$ for any finitely generated free abelian group A. Consequently, $\mathrm{Nil}(\mathbb{Z}\pi_1 M)\otimes\mathbb{Q}$, $\tilde{K}_0(\mathbb{Z}\pi_1 M)\otimes\mathbb{Q}$ and $K_n(\mathbb{Z}\pi_1 M)\otimes\mathbb{Q}$ (where $n < 0$) vanish.*

REMARK. It is now known that $\mathrm{Wh}\,\pi_1 M = 0$ [Farrell and Jones, **1990**]. Chapter 6 contains a discussion of this and other recent results.

The proof of 4.6 follows the format of that of 3.1. In particular, there are three key ingredients:

1. the asymptotic transfer,
2. the geodesic flow, and
3. a foliated control theorem.

For the time being, we drop the assumption that M is a closed manifold. Several of the notions discussed in Chapter 3 for hyperbolic manifolds carry over to this more general situation. The asymptotic transfer of a path $\alpha : I \to M$ to a path $v\alpha$ in SM is defined as before, where $p : SM \to M$ again denotes the tangent unit sphere bundle of M, and $v \in p^{-1}(\alpha(0))$. We still let γ_v be the unique geodesic in M with $\dot\gamma(0) = v$ where $v \in SM$. The definitions of asymptotic geodesic rays and asymptotic vectors as well as of the asymptotic vector field $v(x)$ on X, determined by $v \in SX$, are the same as in Chapter 3. Eberlein and O'Neill [**1973**] construct a compactification $\overline{X}$ of X with many of the properties of $\mathbb{B}^n$ in the Poincaré model for $\mathbb{H}^n$. The pair $(\overline{X}, X(\infty))$ is homeomorphic to $(\mathbb{B}^n, S^{n-1})$ where $X(\infty) = \overline{X} - X$ and

$n = \dim X$. The points of $X(\infty)$ are the asymptotic vector fields on X and each element in Iso X uniquely extends to a homeomorphism of $\overline{X}$.

There is an $\text{Iso}(X)$ equivariant homeomorphism of SX to $X \times X(\infty)$ given by

$$v \mapsto (p(v), v(\ ))$$

where $v \in SX$ and $\text{Iso}(X)$ acts diagonally on $X \times X(\infty)$. Under this identification, the geodesic flow on SX corresponds to the flow g^t on $X \times X(\infty)$ defined by

$$g^t(x, v(\ )) = (\gamma_{v(x)}(t), v(\ ))$$

where $t \in \mathbb{R}$, $x \in X$ and $v(\ ) \in X(\infty)$.

The $\text{Iso}(X)$-invariant subspaces of $X \times X(\infty)$ have the form $X \times F$ where F is an $\text{Iso}(X)$-invariant subspace of $X(\infty)$. Consequently, any $\text{Iso}(X)$-invariant subspace Y of SX is also g^t-invariant as well as being closed under the asymptotic transfer; i.e., $v\alpha$ is a path in Y provided $v \in Y$, $p(v) = \alpha(0)$ and α is a path in X. As in the proof of 3.1, we search for a subspace F of $X(\infty)$ such that $\chi(F) \neq 0$. Note that $F = X(\infty)$ has this property when $\dim X$ is odd. Unfortunately there is another problem; namely, g^t is not an Anosov flow when rank $X > 1$, although it is weakly Anosov. A weakly stable leaf of SX is the image of an asymptotic vector field $v(\ )$. Denote this leaf by $X_{v(\ )}$. Consequently these leaves are parametrized by $X(\infty)$. The curve $v\alpha$ gets no longer when flowed by g^t through positive time. However, it does not become arbitrarily close to a segment (of bounded length) of a flow line (except when rank $X = 1$). It does become "skinny" in a weaker sense to be presently made precise.

4.7. DEFINITION. A pair of vectors v_1, $v_2 \in SX$ are *parallel* (denoted $v_1 \| v_2$) provided both v_1 is asymptotic to v_2 and $-v_1$ is asymptotic to $-v_2$.

Parallelism induces a foliation of $X_{v(\ )}$ whose leaves are equivalence classes of parallel vectors. These leaves are called the *stable leaves of the weakly stable leaf $X_{v(\ )}$*.

4.8. DEFINITION. A path γ in a foliated Riemannian manifold N is (β, ϵ)-*controlled* provided there exists a second path φ in N satisfying the following properties.

(1) The image of φ is contained in a leaf of N and its diameter, measured in this leaf, is less than β.

(2) $d(\gamma(t), \varphi(t)) < \epsilon$ for all $t \in [0, 1]$.

REMARK. This definition agrees with 3.2 when $N = SM$ and is foliated by the flow lines of g^t.

Let $q: X \to M$ denote the covering projection. The images of the weakly stable leaves of SX, under $dq: SX \to SM$, foliate SM and are called its *weakly stable leaves*. The image of the stable foliation of a weakly stable leaf $X_{v(\ )}$ of SX is the *stable foliation* of $dq(X_{v(\ )})$.

4.9. KEY PROPERTY OF $v\alpha$ (a generalization of 3.3). Given a locally symmetric space of noncompact type M, a vector $v \in SM$, and positive real numbers β and ϵ, there exist positive constants s_0 and C such that the following is true. Let α be any smooth path in M whose arc length is less than β and such that $p(v) = \alpha(0)$. Then the composite $g^s \circ (v\alpha)$ is $(C\beta, \epsilon)$-controlled relative to the stable foliation of the weakly stable leaf containing $v\alpha$, provided $s \geq s_0$. Furthermore, C depends only on M and not on v, β or ε.

There is a general foliated control theorem, 3.4 is a special case of it, proven in Farrell and Jones [1988b]. But it requires (in its most easily applicable form) that the leaves of the foliation are flat manifolds in the Riemannian metric induced from the ambient space. We consequently restrict our attention to $\mathrm{Iso}(X)$-invariant subspaces F of $X(\infty)$ satisfying the following two properties.

4.10. REQUIRED PROPERTIES OF F.

(1) The leaves of stable foliation of $X_{v(\ )}$ are flat, for each point $v(\ ) \in F$.

(2) $\chi(F) \neq 0$.

Let G denote the semi-simple linear Lie group $\mathrm{Iso}(X)$ and G^x denote the isotopy subgroup fixing the point $x \in \overline{X}$. If $x \in X$, then G^x is a maximal compact subgroup of G. In fact, X is the space of all maximal compact subgroups in G. If $x \in X(\infty)$, then G^x is a parabolic subgroup of G and every parabolic subgroup occurs in this way. Put more precisely, $X(\infty)$ is the Tits building $\mathscr{T}$ of G retopologized. Recall $\mathscr{T}$ is a simplicial complex whose simplices are the parabolic subgroups of G. If P_1 and P_2 are two parabolic subgroups, then P_1 is a face of P_2 if and only if $P_2 \subset P_1$. (See Mostow [1973, §16].) There is a G-equivariant continuous bijection from $\mathscr{T}$ to $X(\infty)$ under which the simplex corresponding to a parabolic subgroup P is sent to the set all points x in X such that $G^x = P$. This bijection is not a homeomorphism since $\dim \mathscr{T} = r - 1$ while $\dim X(\infty) = n - 1$. (See Ballman, Gromov and Schroeder [1985, Appendix 5] for more details.)

To a parabolic subgroup P of G is associated a reductive group $\mathscr{L}$ where $\mathscr{L}$ is the quotient of P by its unipotent radical $\mathscr{N}$. The group $\mathscr{L}$ is called the (abstract) Levi subgroup of P and P is isomorphic to a semidirect product $\mathscr{L} \ltimes N$. The Levi subgroup is locally the direct product $\mathscr{A} \times \mathscr{S}$ where $\mathscr{A}$ is an abelian group and $\mathscr{S}$ is a semisimple group. The group $\mathscr{A}$ is isomorphic to $\mathbb{R}^s \times \mathscr{C}$ where $\mathscr{C}$ is the maximal compact subgroup of $\mathscr{A}$ and

$$s + (\mathbb{R}\text{-split rank of } \mathscr{S}) = r.$$

If $P = G^{v(\ )}$, then each stable leaf of $X_{v(\ )}$ is isometrically equivalent to a product $\mathbb{E}^s \times X'$ where $\mathbb{E}^s$ is Euclidean s-dimensional space and X' is a symmetric space of noncompact type with

$$\text{rank } X' = \mathbb{R}\text{-split rank } \mathcal{S}.$$

Consequently, the stable leaves of $X_{v(\ )}$ are flat if and only if $G^{v(\ )}$ is a minimal parabolic subgroup G. (Note that the minimal parabolic subgroups of G are all conjugate.)

4.11. EXAMPLE. Let Y_n denote the homogeneous space $\text{SL}_n(\mathbb{C})/\text{SU}_n$ equipped with a $\text{SL}_n(\mathbb{C})$-invariant Riemannian metric (uniquely defined up to a positive constant multiple) where $n \geq 2$. Note that Y_2 is $\mathbb{H}^3$ if the constant multiple is correctly chosen. In general, $\text{rank } Y_n = n - 1$, $\dim Y_n = n^2 - 1$ and X_n is a totally geodesic subspace of Y_n corresponding to the obvious embedding of $\text{SL}_n(\mathbb{R})$ into $\text{SL}_n(\mathbb{C})$. The parabolic subgroups of $\text{SL}_n(\mathbb{C})$ are represented, up to conjugacy, by the different subgroups of blocked upper triangular matrices. Let $T_n(\mathbb{C})$ denote the subgroup consisting of all *upper triangular matrices* in $\text{SL}_n(\mathbb{C})$, then $T_n(\mathbb{C})$ is a minimal parabolic subgroup. Let $\mathcal{N}_n(\mathbb{C})$ denote the subgroup of $T_n(\mathbb{C})$ consisting of all *strictly upper triangular* matrices (i.e., the diagonal entries are all equal to 1), then $\mathcal{N}_n(\mathbb{C})$ is the unipotent radical of $T_n(\mathbb{C})$. The group of all diagonal matrices in $T_n(\mathbb{C})$, is a Levi subgroup of $T_n(\mathbb{C})$. Denote this group by $\mathcal{L}_n(\mathbb{C})$. Since $\mathcal{L}_n(\mathbb{C})$ is abelian, there is no semisimple factor $\mathcal{S}$ in its local decomposition as $\mathcal{A} \times \mathcal{S}$. Pick a vector v_n in SY_n such that $\text{SL}_n(\mathbb{C})^{v_n(\ )} = T_n(\mathbb{C})$ and $p(v_n) = [I]$. The leaf of the stable foliation of $(Y_n)_{v(\ )}$ passing through v_n is the orbit of v_n under the action of $\mathcal{L}_n(\mathbb{C})$ and is isometrically equivalent to $\mathbb{E}^{n-1}$.

Let F_n be the orbit of $v_n(\ ) \in Y_n(\infty)$ under the action of $\text{SL}_n(\mathbb{C})$. Note that F_n is homeomorphic to the homogeneous space $\text{SL}_n(\mathbb{C})/T_n(\mathbb{C})$ which is the manifold consisting of all *flags* in $\mathbb{C}^n$. Since F_n fibers over $\mathbb{C}P^{n-1}$ with fiber F_{n-1}

$$\chi(F_n) = \chi(\mathbb{C}P^{n-1})\chi(F_{n-1}).$$

We obtain from this formula that $\chi(F_n) = n!$ because $\chi(\mathbb{C}P^{n-1}) = n$ and F_1 is a point. The stabilizer of any point in F_n is a minimal parabolic subgroup of $\text{SL}_n(\mathbb{C})$ since it is conjugate to $\text{SL}_n(\mathbb{C})^{v_n(\ )}$. The following result is thus established.

4.12. LEMMA. *The closed* $\text{SL}_n(\mathbb{C})$-*invariant submanifold* F_n *of* $Y_n(\infty)$ *satisfies* 4.10.

We now construct the setup which allows the proof of 3.1 to be modified so as to yield 4.6. Identify $G = \text{Iso}(X)$ with a selfadjoint closed subgroup of $\text{SL}_m(\mathbb{R})$ and thus with a subgroup of $\text{SL}_m(\mathbb{C})$; cf. Mostow [**1973**, §3]. In this way, X and $M = \Gamma \backslash X$ are identified with totally geodesic submanifolds of Y and N, respectively, where Y is an abbreviation for Y_m and N denotes

$\Gamma \backslash Y$. Although N is not compact, its injectivity radius is greater than zero since orthogonal projection of Y onto X is both Γ-equivariant and distance nonincreasing, and M is compact. Let FN denote the balanced product $Y \times_\Gamma F$, where F is an abbreviation for F_m, and $p: FN \to N$ is the bundle projection induced from projecting $Y \times F$ onto its first factor. The proof of 3.1 can now be translated into a proof for 4.6.

REMARKS. The number s_0, posited in 4.9, depends only on β and ϵ provided $v(\ ) \in F$. The application of the (β, ϵ)-Foliated Control Theorem, whose pseudoisotopy version is proven in Farrell and Jones [1988b], requires special attention to the stable leaves whose injectivity radius is not significantly larger than β. We must stratify the union of these, by making systematic use of methods from Mostow [1973], and proceed inductively a stratum at a time; cf, Farrell and Jones [1987b].

An analogue of 4.6 is true for the higher Whitehead groups.

4.13. THEOREM (Farrell and Jones [1987b]). *Let Γ be a discrete, torsion-free, cocompact subgroup of a virtually connected Lie group G. Then $\mathrm{Wh}_n(\Gamma) \otimes \mathbb{Q} = 0$ for all integers n. Consequently,*

$$K_n(\mathbb{Z}\Gamma) \otimes \mathbb{Q} \simeq H_n(\Gamma, \mathbb{Q}) \oplus \left(\bigoplus_{i=1}^{\infty} H_{(n-1)-4i}(\Gamma, \mathbb{Q}) \right)$$

for all integers n.

REMARKS. The proof of 4.6 requires that G is semisimple and linear. The general case, of $\mathrm{Wh} \, \Gamma \otimes \mathbb{Q} = 0$, uses a fibered and F-equivariant (F-finite) form of the above argument following the pattern established in Farrell and Hsiang [1984] and [1987]. To show $\mathrm{Wh}_n(\Gamma) \otimes \mathbb{Q} = 0$ when $n > 1$, h-cobordisms are replaced by pseudoisotopies as in the proof of 3.21. It is also necessary that $FN \to N$ is a topological trivial bundle since the analogue of 3.5 for pseudoisotopies, proved by Hatcher [1978], requires this. This is arranged by changing the original embedding $\rho: G \to \mathrm{SL}_m(\mathbb{R})$ as follows. Note that $FN \to N$ is associated to the linear bundle $Y \times_\Gamma \mathbb{R}^m \to N$, where Γ acts on $\mathbb{R}^m$ via ρ and the bundle projection is induced by mapping $Y \times \mathbb{R}^m$ onto its first factor. Denote this bundle by η. Milnor [1958] shows that the rational Pontryagin classes $p_i(\eta) = 0$, for all $i > 0$, since η is a flat linear bundle. Consequently, there is a positive integer s such that the s-fold direct sum $s\eta = \eta \oplus \eta \oplus \cdots \oplus \eta$ is a trivial linear bundle. Let G act diagonally on $\mathbb{R}^{sm} = \mathbb{R}^m \times \mathbb{R}^m \times \cdots \times \mathbb{R}^m$, where there are s-factors. This gives an embedding $\overline{\rho}: G \to \mathrm{SL}_{sm}(\mathbb{R})$. When ρ is replaced by $\overline{\rho}$, the bundle $FN \to N$ becomes topologically trivial.

5. Existence of Hyperbolic Structures

A hyperbolic structure on a m-dimensional manifold M is a maximal atlas (U_i, f_i) where changes of charts $f_i \circ f_j^{-1}$ are restrictions of isometries of hyperbolic m-space $\mathbb{H}^m$ on each overlap component. Two structures are equivalent provided there is an isometry between them. Mostow's Rigidity Theorem 1.25 shows that a closed manifold M can support at most one equivalence class of hyperbolic structures provided $\dim M \geq 3$.

It is natural to try to topologically characterize those closed manifolds which support a hyperbolic structure. This is easily done for two-dimensional manifolds. Namely, a closed two-dimensional manifold supports a hyperbolic structure if and only if its Euler characteristic is negative. For three-dimensional manifolds, this problem was extensively studied by Thurston [1982].

5.1. THURSTON'S GEOMETRIZATION CONJECTURE. A closed connected orientable three-dimensional manifold M supports a hyperbolic structure if and only if

1. M is aspherical, and
2. each abelian subgroup of $\pi_1 M$ is cyclic.

REMARKS. The necessity of conditions 1 and 2 is classical but it is unknown whether these conditions are sufficient. Conjecture 5.1 implies Poincaré's conjecture 1.3 in the same way that Borel's conjecture 1.22 implies 1.3. Namely, let Σ be a simply connected, closed three-dimensional manifold and N be a closed orientable hyperbolic three-manifold. Then the connected sum $M = N\#\Sigma$ satisfies both conditions 1 and 2 of 5.1. If 5.1 is true, then M supports a hyperbolic structure. Hence Mostow's Rigidity Theorem 1.25 implies M and N are homeomorphic. Consequently, Σ is homeomorphic to S^3 because of Milnor [1962].

There are additional topological conditions on M which guarantee the existence of a hyperbolic structure but these conditions are in general unnecessary.

5.2. THEOREM (Thurston [1982]). *Let M be a closed connected orientable three-manifold which is irreducible and sufficiently large and satisfies condition 2 of 5.1, then M supports a hyperbolic structure.*

REMARK. The conditions irreducible and sufficiently large imply that the orientable manifold M is aspherical; i.e., satisfies condition 1 of 5.1.

The next result solves the problem for high dimensional manifolds.

5.3. THEOREM (Farrell and Jones [1989]). *A closed connected m-dimensional manifold M (where $m > 4$) supports a hyperbolic structure if and only if*

1. *M is aspherical, and*
2. *$\pi_1 M$ is isomorphic to a discrete cocompact subgroup of $O(m, 1)$.*

REMARKS. It is a classical result that conditions 1 and 2 are necessary, even when $m < 5$. It is unknown whether these conditions are sufficient when $m = 4$.

Theorem 5.3 is a direct consequence of the following result.

5.4. THEOREM. (Farrell and Jones [1989]). *Let N be a closed hyperbolic manifold with $\dim N > 4$. Then $|\mathscr{S}(N)| = 1$; i.e., any homotopy equivalence $f: M \to N$, where M is a closed manifold, is homotopic to a homeomorphism.*

REMARKS. Mishchenko [1974] proved that $f^*(p_i N) = p_i M$, for all i, where p_i is ith rational Pontryagin class. Farrell and Hsiang [1981] showed that $f \times \mathrm{id}: M \times \mathbb{R}^3 \to N \times \mathbb{R}^3$ is properly homotopic to a homeomorphism. When N is a (orientable) sufficiently large three-manifold, Cappell [1976] calculated the surgery obstruction groups $L_i(\pi_1 N) \otimes \mathbb{Z}[\frac{1}{2}]$. Nicas and Stark [1984], [1987] and Nicas [1986] calculated $L_i(\pi_1 N) \otimes \mathbb{Z}[\frac{1}{2}]$ for certain interesting classes of nonsufficiently large (orientable) hyperbolic three-dimensional manifolds N.

DERIVATION OF 5.3 FROM 5.4. Let N be the double coset space $\Gamma \backslash O(m, 1)/(O(m) \times O(1))$ where Γ is a discrete cocompact subgroup of $O(m, 1)$ which is isomorphic to $\pi_1 M$. The closed manifold N inherits a hyperbolic structure from a $O(m, 1)$-invariant Riemannian metric on $O(m, 1)/(O(m) \times O(1))$. Since both M and N are aspherical and have isomorphic fundamental groups, they are homotopy equivalent and hence homeomorphic because of 5.4. Pull the hyperbolic structure on N back to M via this homeomorphism. □

The proof of 5.4 is indirect via a transfer argument. Surgery theory shows that f is homotopic to a homeomorphism provided $f \times \mathrm{id}: M \times \mathbb{C}P^n \to N \times \mathbb{C}P^n$ is homotopic to a homeomorphism for some even integer n. (See Wall [1971] and Kirby and Siebenmann [1977] for an extensive discussion of surgery theory.) A more complicated transfer reduction is actually used. It is formulated as follows. For each positive integer n, F_n denotes the space consisting of all unordered pairs $[u, v]$ of vectors u, v in S^{n-1}, i.e.,

$$F_n = (S^{n-1} \times S^{n-1})/C_2$$

where the cyclic group C_2 (of order 2) acts on $S^{n-1} \times S^{n-1}$ by interchanging coordinates. (Notice that $F_2 = \mathbb{C}P^2$). Since S^{n-1} is the boundary of $\mathbb{H}^n$ in the Poincaré model, $\mathrm{Iso}\,\mathbb{H}^n$ acts diagonally on $S^{n-1} \times S^{n-1}$ and this action commutes with that of C_2. Therefore, the diagonal action of $\mathrm{Iso}\,\mathbb{H}^n$ on $S^{n-1} \times S^{n-1}$ induces an action of $\mathrm{Iso}\,\mathbb{H}^n$ on F_n. There is a natural stratification of F_n consisting of two strata B and T. The bottom stratum B consists of all (agreeing) unordered pairs $[u, v]$ where $u = v$, while the top stratum T consists of all (disagreeing) pairs $[u, v]$ where $u \neq v$. The action of $\mathrm{Iso}\,\mathbb{H}^n$ on F_n respects this stratification. Identify N with an orbit space $\mathbb{H}^n/\Gamma$ where Γ is a torsion-free discrete subgroup of $\mathrm{Iso}\,\mathbb{H}^n$. Let FN, $\mathbb{B}$ and $\mathbb{T}$ denote the orbit spaces of $\mathbb{H}^n \times F_n$, $\mathbb{H}^n \times B$ and $\mathbb{H}^n \times T$, respectively, under the diagonal action of Γ. Note that FN is a stratified space with two strata $\mathbb{B}$ and $\mathbb{T}$.

Projection onto the first factor of $\mathbb{H}^n \times F_n$ induces a fiber bundle $FN \to N$ with fiber F_n. Let E denote the total space of the pullback of this bundle along f, and $f^* \colon E \to FN$ be the canonical map covering f.

5.5. PROPOSITION. *If* $\dim N$ *is an odd integer greater than* 3, *then the following is true. The map* f *is homotopic to a homeomorphism provided* f^* *is admissibly homotopic to a map* g *satisfying the following properties. The map* g *is a homeomorphism over* $\mathbb{B}$ *and over the complement of a tubular neighborhood for the stratum* $\mathbb{B}$ *in* FN, *as well as being "blocked up" over this tubular neighborhood (i.e., split with respect to some triangulation of* $\mathbb{B}$).

REMARK. An *admissible* homotopy is strata preserving and a bundle map homotopy in a tubular neighborhood of the bottom stratum.

This result is plausible because of the following facts about F_n.

5.6. LEMMA. *If* n *is any odd integer greater than* 1, *then* F_n *has the following properties*:

1. F_n *is an orientable,* $2n - 2$ *dimensional* $\mathbb{Z}[\tfrac{1}{2}]$-*homology manifold.*
2. F_n *is simply connected.*

3.
$$H_i(F_n) = \begin{cases} \mathbb{Z} & \text{if } i = 0,\, n-1,\ \text{or } 2n-2 \\ \mathbb{Z}_2 & \text{if } n \leq i < 2n-2 \text{ and } i \text{ is even} \\ 0 & \text{otherwise} \end{cases}$$

4.
$$H^i(F_n) = \begin{cases} \mathbb{Z} & \text{if } i = 0,\, n-1,\ \text{or } 2n-2 \\ \mathbb{Z}_2 & \text{if } n < i < 2n-2 \text{ and } i \text{ is odd} \\ 0 & \text{otherwise} \end{cases}$$

5. *The cup product pairing* $H^{n-1}(F_n) \otimes H^{n-1}(F_n) \to H^{2n-2}(F_n)$ *is unimodular with signature* ± 1.

REMARK. $H_i(F_n)$ and $H^i(F_n)$ denote the i-dimensional homology and cohomology groups of F_n with untwisted $\mathbb{Z}$-coefficients.

It is seen that the space F_n is the union of the total spaces of two bundles. One is a bundle over S^{n-1} with fiber the cone on P^{n-2}. In fact, it is a tubular neighborhood of $B = S^{n-1}$ in F_n. The other is a bundle over P^{n-1} with fiber $\mathbb{D}^{n-1}$. The second bundle is a tubular neighborhood of P^{n-1} in T where P^{n-1} is identified with the set of all unordered pairs $[u, -u]$ in F_n. The two sets intersect in the total space of the P^{n-2} bundle associated to the tangent bundle of S^{n-1}. This description of F_n is useful in verifying the first four properties of 5.6. Property five can be seen by studying the following diagram, in which a and b are distinct points in S^{n-1}, and observing that the homology classes represented by $a \times S^{n-1}$ and $S^{n-1} \times b$ both generate $H_{n-1}(F_n)$.

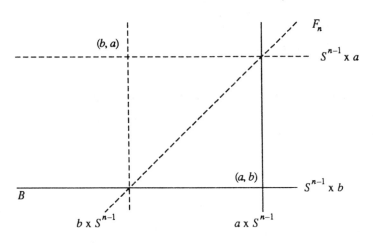

FIGURE 4

To prove 5.4, at least when $n = \dim N$ is an odd integer greater than 5, it suffices to construct an admissible homotopy from f^* to a map g with the properties posited in 5.5. This construction uses extra structure which the strata of FN possess. There is a foliation $\mathscr{F}$ of FN whose leaves are the images of the sets $\mathbb{H}^n \times a$, $a \in F_n$, under the covering projection $q : \mathbb{H}^n \times F_n \to FN$. This foliation restricts to foliations $\mathscr{F}_\mathbb{B}$ and $\mathscr{F}_\mathbb{T}$ for the strata $\mathbb{B}$ and $\mathbb{T}$, respectively.

The geometry of N gives rise to additional structures, called *markings*, on the leaves $q(\mathbb{H}^n \times a)$ of $\mathscr{F}_\mathbb{B}$ and $\mathscr{F}_\mathbb{T}$. Each leaf of $\mathscr{F}_\mathbb{B}$ is marked with an *asymptotic* vector field and each leaf of $\mathscr{F}_\mathbb{T}$ is marked with a geodesic. Recall, from Chapter 3, that the points on S^{n-1} are identified with the set of *asymptotic* vector fields on $\mathbb{H}^n$; i.e., vector fields of the form $v(\ )$ where $v \in S\mathbb{H}^n$. The leaf $L = q(\mathbb{H}^n \times a)$ is contained in $\mathbb{B}$ if and only if $a \in B$. In this case, there exists a vector $v \in S\mathbb{H}^n$ such that $a = [v(\ ), v(\ )]$ and L is marked with the vector field $dq \circ v(\ )$. That is, the valve of this vector field at the point $q(x, a)$ in L is $dq(v(x))$. The leaf L is in $\mathbb{T}$ if and only if $a \in T$; i.e., $a = [u, v]$ where u and v are distinct points in S^{n-1}. There exists a unique geodesic line l in $\mathbb{H}^n$ connecting u and v; i.e., such that the closure of l in $\mathbb{R}^n$ contains both u and v. (Recall that a geodesic line is the intersection with $\mathbb{H}^n$ of either a circle or straight line in $\mathbb{R}^n$ which meets S^{n-1} perpendicularly.) Then $q(l)$ marks L.

The union of the markings of the leaves of $\mathscr{F}_\mathbb{T}$ is called the *core* C of the top stratum $\mathbb{T}$. The spaces $\mathbb{B}$ and C can be identified with SN and PN, respectively, where PN denotes the total space of the P^{n-1}-bundle associated to $SN \to N$. The identifications are induced by the following pair of Γ-equivariant maps

$$\varphi \colon S\mathbb{H}^n \to \mathbb{H}^n \times B \quad \text{and}$$
$$\psi \colon P\mathbb{H}^n \to \mathbb{H}^n \times T.$$

Let $r \colon S\mathbb{H}^n \to \mathbb{H}^n$ denote the bundle projection, then

$$\varphi(v) = (r(v), v(\ ))$$

where $v \in S\mathbb{H}^n$. Note that a point in $P\mathbb{H}^n$ is an unordered pair of vectors $[v, -v]$ where $v \in S\mathbb{H}^n$. Recall, from Chapter 3, that a vector $v \in S\mathbb{H}^n$ determines a geodesic α_v in $\mathbb{H}^n$. Let $\alpha_v(+\infty)$ and $\alpha_v(-\infty)$ be the points on S^{n-1} defined by

$$\alpha_v(+\infty) = \lim_{t \to +\infty} \alpha_v(t) \quad \text{and}$$
$$\alpha_v(-\infty) = \lim_{t \to -\infty} \alpha_v(t)$$

where these limits are taken relative to the Euclidean metric on $\mathbb{R}^n$. The map ψ is defined by

$$\psi([v, -v]) = (r(v), [\alpha_v(+\infty), \alpha_v(-\infty)])$$

where $v \in S\mathbb{H}^n$.

Integrating the markings on the leaves of $\mathscr{F}_\mathbb{B}$ determines a flow on $\mathbb{B}$. This flow is the geodesic flow g^t on SN under the identification of $\mathbb{B}$ and SN. The flow lines of g^t foliate SN and their images, under the two-sheeted covering projection $SN \to PN$, give a one-dimensional foliation $\mathscr{G}$ to PN. The leaves of $\mathscr{G}$ become the markings of the leaves of $\mathscr{F}_\mathbb{T}$ under the identification of PN with C.

There is a fiber bundle $p \colon \mathbb{T} \to C$ with fiber $\mathbb{R}^{n-1}$ such that p restricted to a leaf L of $\mathscr{F}_\mathbb{T}$ is orthogonal projection onto its marking $q(l)$. (Here orthogonal projection is with respect to the *leaf metric* on $L = q(\mathbb{H}^n \times a)$ which is a hyperbolic structure on L, in its leaf topology, induced by the covering projection $q \colon \mathbb{H}^n = \mathbb{H}^n \times a \to L$.)

There is a vector field $w(\ )$ on $\mathbb{T} - C$ tangent to the leaves of the foliation $\mathscr{F}_\mathbb{T}$ and pointing toward C. More explicitly, let L be the leaf of $\mathscr{F}_\mathbb{T}$

containing a point $x \in \mathbb{T} - C$. Connect x to $p(x)$ via a unit speed geodesic in L, using the leaf metric on L. Then $w(x)$ is the tangent vector to this geodesic at x. This vector field integrates to give an incomplete *radial flow* r^t on $\mathbb{T}$. In particular, $r^t(x)$ is only defined for $t \in [0, d_L(x, p(x))]$, where $d_L(x, y)$ denotes the distance between two points $x, y \in L$ using the leaf metric on L.   ∘

Given a path α in N and a point $w = [u, v]$ in FN with $r(w) = \alpha(0)$, there is an asymptotic transfer $w\alpha$. (Here r denotes the bundle projection $F\dot{N} \to N$.) This is the path in FN defined by

$$w\alpha(t) = [u\alpha(t), v\alpha(t)]$$

where $u\alpha$ and $v\alpha$ are asymptotic transfers of α to SN defined in Chapter 3.

REMARK. Let w be any point in $\mathbb{T}$ and α be any path in N such that $r(w) = \alpha(0)$. Then $p \circ \alpha w$ is contained in a leaf of the foliation $\mathscr{G}$ of C.

A *homotopy inverse structure* for $f: M \to N$ is a map $\varphi: N \to M$ together with homotopies h_t from $f \circ \varphi$ to id_N and k_t from $\varphi \circ f$ to id_M. A structure determines a collection of paths in N called its *tracks*. The tracks are parameterized by $N \cup M$ and consist of all the paths

$$t \mapsto h_t(x) \quad \text{and}$$
$$t \mapsto f(k_t(y))$$

where $x \in N$ and $y \in M$. The homotopy equivalence is ϵ-*controlled* if it has a homotopy inverse structure such that all of its tracks have diameter less than ϵ.

5.7. THEOREM (Chapman and Ferry [1979]). *Given a closed Riemannian manifold N, with $\dim N \geq 5$, there exists a number $\epsilon > 0$ such that any ϵ-controlled homotopy equivalence $f: M \to N$, where M is a closed manifold, is homotopic to a homeomorphism.*

REMARKS. There is a version of 5.7, relative to a closed subset K of N, which implies f is homotopic to a homeomorphism over K provided it is ϵ-controlled over a neighborhood of K. In this version, ϵ depends on $K \subset N$. Quinn [1979] proved a fibered version of 5.7 where $p: N \to X$ is a fiber bundle whose fiber is a compact connected manifold and the diameters of the tracks are measured in X. That is, the image of each path $p \circ \gamma$ must have diameter less than ϵ where γ is any track of the homotopy inverse structure. The number ϵ, in this result, depends only on X but it is necessary that $\dim F \geq 5$ and $\mathrm{Wh}(\pi_1(F \times T^m)) = 0$ for all $m \geq 0$. The conclusion is that f is homotopic to a map which is split with respect to a triangulation of X but is not necessarily a homeomorphism. Farrell and Jones [1988c] prove a fibered version of 5.7 with (β, ϵ)-foliated control; cf. 3.2 and 3.4. Here the base space of the bundle is either SN or PN where N is a closed hyperbolic manifold. The flow lines of g^t foliate SN and $\mathscr{G}$ is the foliation of PN. To avoid confusion, $\mathscr{N}$ (instead of N) here denotes the range of f and the domain of ρ. An additional requirement is that

$$\text{Wh}\,\pi_1(\rho^{-1}(\mathscr{O}) \times T^m) = 0$$

for each closed leaf $\mathscr{O}$ in the base and every $m \geq 0$. There is also a relative version where f is already a homeomorphism over $\partial\mathscr{N}$. The conclusion then is that f is homotopic, via a homotopy which is constant over $\partial\mathscr{N}$, to a map which is "blocked up".

It can be assumed, in proving 5.1, that M is a smooth manifold and f is a smooth map. (This is a consequence of Farrell and Hsiang [1981b].) Let φ, h_t and k_t be a homotopy inverse structure for f (denote this structure by Φ) consisting of smooth maps and let $\{\gamma_x | x \in N \cup M\}$ denote the tracks of Φ. It is unlikely that all of these tracks have diameter less than the number ϵ posited in 5.7 relative to N. However there exists a homotopy inverse structure $\overline{\Phi} = \{\overline{\varphi}, \overline{h}_t, \overline{k}_t\}$ for $f^*: E \to FN$ whose tracks are the set of asymptotic transfers $\{w\gamma_x | x \in N \cup M\}$ of the tracks of Φ. Furthermore the map $\overline{\varphi}$ as well as the homotopies $\overline{h}_t$ and $\overline{k}_t$ are strata preserving. The admissible homotopy to a map g satisfying 5.5 is accomplished in three steps; each of which uses either 5.7 or the Remark following 5.7. The first step is to flow the tracks $w\gamma$ of $\overline{\Phi}$, where $w \in \mathbb{B} = SN$, sufficiently far using g^t. The flowed paths are then (β, ϵ)-controlled, relative to the foliation of SN by the flow lines of g^t, where ϵ can be arbitrarily small, and β is fixed. An application of the Remark to 5.7 yields a homotopy of f^* to a homeomorphism over the lower stratum $\mathbb{B}$. The covering homotopy theorem extends this homotopy over a tubular neighborhood $\mathscr{N}_1$ of $\mathbb{B}$.

The second step extends this homotopy over the complement of a tubular neighborhood $\mathscr{N}_2$ of the union $\mathbb{B} \cup C$, where $\mathscr{N}_1$ is contained in the interior of $\mathscr{N}_2$. The radial flow r^t applied to the tracks $w\gamma$ of $\overline{\Phi}$, where $w \in \mathbb{T}$, shrinks their size. The diameters of the images of the flowed paths which come to rest inside (or close to) the complement of $\mathscr{N}_2$ can be made arbitrarily small by appropriately tapering r^t. The tapering is such that any flowed path which meets $FN - \mathscr{N}_2$ must originate very far away from the core C (as measured in the leaf metric); also the time interval must be compressed to guarantee control in the flow line direction. An application of 5.7 yields the admissible homotopy over $FN - \mathscr{N}_2$.

The space $\mathscr{N}_2$ is the disjoint union of tubular neighborhoods $\mathscr{N}_2(C)$ of C and $\mathscr{N}_2(\mathbb{B})$ of $\mathbb{B}$. It can be chosen so that $p: \mathscr{N}_2(C) \to C = PN$ is a fiber bundle whose fiber is a closed $(n-1)$-ball. It can also be arranged that the images, under p, of the "new tracks" of $\overline{\Phi}$ (after completing step 2) are (β, ϵ)-controlled relative to $\mathscr{G}$ for arbitrarily small ϵ, and fixed β. By an application of the Remark to 5.7 and the Remark immediately preceding, together with the fact that $|\mathscr{S}(\mathbb{B}^i, S^{i-1})| = 1$ when $i > 4$, the admissible homotopy of f^* can be extended over $\mathscr{N}_2(C)$ to a homeomorphism. This does half of step 3. (The statement $|\mathscr{S}(\mathbb{B}^i, S^{i-1})| = 1$ means that any homotopy equivalence $\varphi: (\mathscr{M}, \partial\mathscr{M}) \to (\mathbb{B}^i, S^{i-1})$, where $\mathscr{M}$ is a compact manifold and $\varphi: \partial\mathscr{M} \to S^{i-1}$ is a homeomorphism, is homotopic to a homeomorphism via a homotopy which is constant on the boundary. This statement is a consequence of 1.4 and 1.7.)

Let $\operatorname{Int}\mathscr{N}_1$ denote the interior of $\mathscr{N}_1$ and

$$\rho: \mathscr{N}_2(\mathbb{B}) - \operatorname{Int}\mathscr{N}_1 \to \mathbb{B} = SN$$

be the bundle projection. Its fiber is homeomorphic to $P^{n-2} \times [0, 1]$. By picking $\mathscr{N}_2(\mathbb{B})$ small enough and t sufficiently large, it can be arranged that the images, under $g^t \circ \rho$, of the "new tracks" of $\overline{\Phi}$ (those produced by step 2) are (β, ϵ)-controlled, relative to the flow lines of g^t, for arbitrarily small ϵ, and fixed β. Hence by once more applying the Remark to 5.7, the admissible homotopy f^* is extended over the remainder of FN. It is here that g cannot be made a homeomorphism but only "blocked up" since $|\mathscr{S}(P^{n-2} \times I^k, \partial)| > 1$, in general. This completes the discussion of 5.4.

REMARK. The second half of the last step used that $\operatorname{Wh}(\pi_1(P^{n-2} \times T^m)) = 0$ for all n and m. The key to this observation is that there exists a cartesian square of rings (when $n > 3$)

$$\begin{array}{ccc}
\mathbb{Z}\pi_1(P^{n-2} \times T^m) & \to & \mathbb{Z}\pi_1 T^m \oplus \mathbb{Z}\pi_1 T^m \\
\downarrow & & \downarrow \\
\mathbb{Z}_2\pi_1 T^m & \to & \mathbb{Z}_2\pi_1 T^m \oplus \mathbb{Z}_2\pi_1 T^m
\end{array}$$

which yields a "Mayer-Vietoris type" exact sequence

$$K_2\mathbb{Z}_2\pi_1 T^m \oplus K_2\mathbb{Z}_2\pi_1 T^n \to K_1\mathbb{Z}\pi_1(P^{n-2} \times T^m) \to K_1\mathbb{Z}_2\pi_1 T^m \oplus K_1\mathbb{Z}\pi_1 T^m \oplus K_1\mathbb{Z}\pi_1 T^m .$$

The following variant of 5.4 can be used, in conjunction with 3.13, to obtain information about the homeomorphism group of a closed hyperbolic manifold N.

5.8. THEOREM (Farrell and Jones [1989]). *Let N be a closed hyperbolic manifold and $f: M \to N \times \mathbb{B}^k$ be a homotopy equivalence where M is a compact manifold with $k + \dim N > 4$. Assume that f maps ∂M homeomorphically to $N \times S^{k-1}$. Then f is homotopic to a homeomorphism via a homotopy which is constant when restricted to ∂M.*

5.9. DEFINITION. Let M be a closed manifold. Then $\operatorname{Top} M$ denotes the topological group whose elements are the self-homeomorphisms of M, equipped with the compact-open topology. It contains a normal closed subgroup $\mathscr{T}\!op\, M$ consisting of all homeomorphisms which are freely homotopic to id_M. The outer automorphism group $\operatorname{Out} \pi_1 M$ is the quotient of the automorphism group $\operatorname{Aut} \pi_1 M$ by its normal subgroup of all inner automorphisms. The following fact is a consequence of 1.25.

5.10. THEOREM (Mostow [**1967**]). *Let* M *be a closed hyperbolic manifold with* dim $M \geq 3$. *Then* Top M *is the semidirect product* $\mathcal{T}$op $M \rtimes$ Iso M *where* Iso M *denotes the group of all isometries of* M. *Furthermore,* Iso M *is a finite discrete group and is isomorphic to* Out $\pi_1 M$.

Let Top$(S^1 \times \mathbb{B}^{m-1}, \partial)$ denote the topological group of all self-homeomorphisms of $S^1 \times \mathbb{B}^{m-1}$ which leave fixed every point of $\partial(S^1 \times \mathbb{B}^{m-1}) = S^1 \times S^{m-2}$. This space is also equipped with the compact-open topology.

5.11. DEFINITION. Let $\times$ Top$(S^1 \times \mathbb{B}^{m-1}, \partial)$ denote the cartesian product of a countably infinite number of copies of Top$(S^1 \times \mathbb{B}^{m-1}, \partial)$. Equip this product with the box topology. In particular, a point in $\times$ Top$(S^1 \times \mathbb{B}^{m-1}, \partial)$ is a sequence f_i of homeomorphisms in Top$(S^1 \times \mathbb{B}^{m-1}, \partial)$. Then $\sum$ Top$(S^1 \times \mathbb{B}^{m-1}, \partial)$ denotes the subspace of $\times$ Top$(S^1 \times \mathbb{B}^{m-1}, \partial)$ consisting of all sequences f_i such that f_i is the identity homeomorphism except for a finite number of indices i.

REMARKS. A basis for the box topology on $\times$ Top$(S^1 \times \mathbb{B}^{m-1}, \partial)$ consists of all cartesian products $\times U_i$ where U_i is an arbitrary sequence of open sets in Top$(S^1 \times \mathbb{B}^{m-1}, \partial)$. The direct limit as $j \to +\infty$, of the finite j-factor cartesian products

$$\text{Top}(S^1 \times \mathbb{B}^{m-1}, \partial) \times \text{Top}(S^1 \times \mathbb{B}^{m-1}, \partial) \times \cdots \times \text{Top}(S^1 \times \mathbb{B}^{m-1}, \partial)$$

is an alternate description of $\sum$ Top$(S^1 \times \mathbb{B}^{m-1}, \partial)$. The space $\times$ Top$(S^1 \times \mathbb{B}^{m-1}, \partial)$ is a topological group using co-ordinatewise multiplication, and $\sum$ Top$(S^1 \times \mathbb{B}^{m-1}, \partial)$ is a closed subgroup.

5.12. THEOREM (Farrell and Jones [**1989**]). *Suppose* M *is a closed orientable hyperbolic manifold and* $m = $ dim $M > 10$. *Then there is a continuous group monomorphism*

$$\varphi: \sum \text{Top}(S^1 \times \mathbb{B}^{m-1}, \partial) \to \mathcal{T}\text{op } M$$

such that

$$\varphi_{\#}: \pi_i(\sum \text{Top}(S^1 \times \mathbb{B}^{m-1}, \partial)) \to \pi_i \mathcal{T}\text{op } M$$

is an isomorphism for every integer i *in the interval* $0 \leq i \leq (m-7)/3$.

PROOF. The map φ is constructed as follows. Let T_n, $n \in \mathbb{Z}^+$, be a pairwise disjoint sequence of copies of $S^1 \times \mathbb{B}^{m-1}$ embedded in M whose cores represent the set of conjugacy classes of pairs $\{\gamma, \gamma^{-1}\}$ of indivisible elements in $\pi_1 M$. If the sequence f_n of homeomorphisms of $S^1 \times \mathbb{B}^{m-1}$ represents an element f in $\sum$ Top$(S^1 \times \mathbb{B}^{m-1}, \partial)$, then $\varphi(f)$ is the homeomorphism of M such that $\varphi(f)$ restricted to T_n is f_n and is the inclusion map on $M - \bigcup T_n$.

Let $G(M)$ denote the H-space of all self-homotopy equivalences of M equipped with the compact-open topology. Note that the finite discrete space $\text{Iso}\,M$ is a deformation retract of $G(M)$. Hence the homotopy fiber of the map between classifying spaces induced by the inclusion of $\text{Top}\,M$ into $G(M)$ is the classifying space of $\mathscr{T}\text{op}\,M$; i.e., the homotopy fiber of $\text{BTop}\,M \to \text{B}G(M)$ is $\text{B}\mathscr{T}\text{op}\,M$. (These results use 5.10 and the fact that $\pi_1 M$ has trivial center.) The function space interpretation of the surgery exact sequence given by Quinn [**1970**] together with 5.8 show that the first quadrant spectral sequence E^n_{pq}, constructed by Hatcher [**1978, Proposition 2.1**], converges to $\pi_{p+q}\mathscr{T}\text{op}\,M$. (Theorem 5.8 shows that the homotopy fiber of the map $\widetilde{\text{BTop}M} \to \text{B}G(M)$ is contractible, where $\widetilde{\text{Top}M}$ is the simplicial group of block homeomorphisms of M.) This spectral sequence has

$$E^1_{pq} = \pi_q \mathscr{P}(M)$$

provided q is within a "stable range" estimated by Igusa [**1988**]; namely, $q \leq (m-7)/3$. A similar analysis holds for $\text{Top}(S^1 \times \mathbb{B}^{m-1}, \partial)$. (Note $G(S^1 \times \mathbb{B}^{m-1}, \partial)$ is contractible.)

A first quadrant spectral sequence $\mathscr{E}^n_{pq}$ and a map between spectral sequences $\Phi^n_{pq}: \mathscr{E}^n_{pq} \to E^n_{pq}$, with the following properties, can be constructed in this way:

1. $\mathscr{E}^1_{pq} = \pi_q \overline{\mathscr{P}}(S^1)$ provided $q \leq (m-7)/3$;
2. $\mathscr{E}^n_{pq}$ converges to $\pi_{p+q}(\sum \text{Top}(S^1 \times \mathbb{B}^{m-1}, \partial))$;
3. Φ^n_{pq} converges to
   $$\varphi_\#: \pi_{p+q}(\sum \text{Top}(S^1 \times \mathbb{B}^{m-1}, \partial)) \to \pi_{p+q}\mathscr{T}\text{op}\,M$$

Property 1 together with 3.14 and 3.12 show that Φ^1_{pq} is an isomorphism provided that $q \leq (m-7)/3$. Consequently, a comparison theorem (see Farrell and Jones [**1989, Lemma 10.10**]) shows that Φ^n_{pq} is an isomorphism for all $n \geq 1$ (including $n = \infty$) provided $p + q \leq (m-7)/3$. This fact together with property 3 complete the proof of 5.12. $\square$

Information about $\text{Top}(S^1 \times \mathbb{B}^{m-1}, \partial)$ can be combined with 5.12 to produce information about $\text{Top}\,M$. Here are two important facts.

5.13. THEOREM (Hatcher [**1978**]). *The abelian group* $\pi_0 \text{Top}(S^1 \times \mathbb{B}^{m-1}, \partial)$ *is countably infinite and has exponent* 2, *provided* $m \geq 6$. *That is, it is the direct sum of a countably infinite number of copies of the cyclic group of order two.*

The second fact is a consequence of 3.16 by using Hatcher's spectral sequence cited above.

5.14. THEOREM. *The vector space* $\pi_i \text{Top}(S^1 \times \mathbb{B}^{m-1}, \partial) \otimes \mathbb{Q} = 0$, *provided* $i \leq (m-7)/3$.

The next result is a consequence of combining 5.12 with 5.13 and 5.14.

5.15. COROLLARY (Farrell and Jones [**1989**]). *Let M be a closed hyperbolic manifold such that $m = \dim M > 10$. Then $\pi_i \operatorname{Top} M \otimes \mathbb{Q} = 0$ provided $1 \le i \le (m-7)/3$, and $\pi_0 \operatorname{Top} M$ is the semidirect product $\mathbb{Z}_2^\infty \rtimes \operatorname{Iso} M$ where $\mathbb{Z}_2^\infty$ denotes the countably infinite group of exponent 2 and $\operatorname{Iso} M$ is the group of all isometries of M. (Recall that $\operatorname{Iso} M$ is a finite group and is isomorphic to $\operatorname{Out} \pi_1 M$.)*

REMARK. The proof of 5.15 requires some extra reasoning when M is nonorientable.

Let $\operatorname{Diff} M$ denote the group of all diffeomorphisms of the closed smooth manifold M.

5.16. COROLLARY (Farrell and Jones [**1989**]). *Let M be a closed hyperbolic manifold with $m = \dim M > 10$, and s be any integer satisfying $1 \le s \le (m - 7)/3$. Then*

$$\pi_s \operatorname{Diff} M \otimes \mathbb{Q} = \begin{cases} \bigoplus_{j=1}^\infty H_{(s+1)-4j}(M, \mathbb{Q}), & \text{if } m \text{ is odd} \\ 0, & \text{if } n \text{ is even.} \end{cases}$$

PROOF. Farrell and Hsiang [**1978b, Theorem 4.5**] reduced this result to 3.21 and 5.8 but with a stricter constraint on s. However, 3.12 allows this constraint to be relaxed to $1 \le s \le (m - 7)/3$. □

6. Epilogue

This chapter contains a brief account of extensions (obtained since the CBMS conference) of many of the results from earlier chapters to the class of nonpositively curved manifolds. In addition, there is a discussion of some constructions of examples of nonclassical aspherical manifolds. Throughout this chapter, M will denote a compact connected (closed) Riemannian manifold all of whose sectional curvature values are nonpositive. We start by stating a generalization of 5.4.

6.1. THEOREM (Farrell and Jones [**1990**]). *Let* $f: N \to M$ *be a homotopy equivalence where* N *is a compact manifold and assume* $\dim M > 4$. *Then* f *is homotopic to a homeomorphism* (*i.e.,* $|\mathscr{S}(M)| = 1$).

This leads to the following topological characterization of compact locally symmetric spaces of noncompact type and of $\dim \neq 3, 4$. It is an existence statement to match Mostow's uniqueness Theorem 1.26.

6.2. COROLLARY (Farrell and Jones). *A closed connected* m-*dimensional manifold* N (*where* $m > 4$) *supports the structure of a locally symmetric space of noncompact type if and only if*

1. N *is aspherical, and*
2. $\pi_1 N$ *is isomorphic to a discrete cocompact subgroup of a* (*virtually connected*) *semisimple Lie group.*

PROOF. The necessity of conditions 1 and 2 is classical; cf., 4.1 and the remarks following 4.4. The sufficiency of these conditions is seen as follows. Let G be the semisimple Lie group containing a cocompact discrete subgroup Γ isomorphic to $\pi_1 N$. Then N and the double coset space $\Gamma \backslash G / K$, denoted by M, are homotopically equivalent where K is a maximal compact subgroup of G. The remarks following 4.4 show that M is a compact locally symmetric space of noncompact type which is nonpositively curved by 4.1. Hence 6.1 shows that N is homeomorphic to M. Pull the locally symmetric space structure back to N via this homeomorphism. $\square$

The next result improves the vanishing result 4.6 by showing it is unnecessary to tensor with $\mathbb{Q}$.

6.3. COROLLARY (Farrell and Jones [1990]). *Let A be any finitely generated free abelian group, including 0. Then $\mathrm{Wh}(\pi_1 M \times A) = 0$. Consequently, $\mathrm{Wh}\,\pi_1 M$, $\tilde{K}_0(\mathbb{Z}\pi_1 M)$, $\mathrm{Nil}(\mathbb{Z}\pi_1 N)$ and $K_n(\mathbb{Z}\pi_1 M)$ (where $n < 0$) all vanish.*

PROOF. Let T^n be a flat torus such that $\pi_1 T^n \simeq A$. Then $M \times T^n$ is a closed nonpositively curved manifold and $\pi_1 M \times T^n \simeq \pi_1 M \times A$. We may assume $n > 4$. Thus it suffices to consider the case where $A = 0$ and $\dim M > 4$. Let $x \in \mathrm{Wh}\,\pi_1 M$ and (W, M) be a h-cobordism such that $\tau(W, M) = x$. Let N be the top of W and $f: N \to M$ be the composite of the inclusion N into W with a retraction of W onto M, then f is homotopic to a homeomorphism $\varphi: N \to M$, because of 5.1. Let $\mathscr{M}$ be the mapping torus of φ. There is a homotopy equivalence $g: \mathscr{M} \to M \times S^1$ such that $\tau(g)$ is $\sigma_\#(x)$, where $\sigma: \pi_1 M \to \pi_1 M \times C$ is the inclusion map onto the first factor. Note that $\sigma_\#$ is monic, because Wh is a functor. Consequently it suffices to show that $\tau(g) = 0$. But g is homotopic to a homeomorphism because of 5.1. Therefore, 1.13 shows that $\tau(g) = 0$. $\square$

Farrell and Jones [1990] show that the weak homotopy type of $\mathscr{P}(M)$ is calculable in terms of $\mathscr{P}(S^1)$ through a stable range of dimensions. This generalizes 3.14. The calculation is in terms of the structure of families of closed geodesics in M. When M is a locally symmetric space of rank > 1, this structure can be complicated. Here are some consequences of this calculation and of a generalization of 6.1 which is analogous to 5.8.

6.4. THEOREM (Farrell and Jones [1990]). *Let n be any positive integer and $\nu = [(n+1)/2]!$, then $\mathrm{Wh}_n(\pi_1 M) \otimes \mathbb{Z}[\frac{1}{\nu}] = 0$ and consequently*

$$K_n(\mathbb{Z}\pi_1 M) \otimes \mathbb{Q} = H_n(M, \mathbb{Q}) \oplus (\oplus_{i=1}^{\infty} H_{n-1-4i}(M, \mathbb{Q})).$$

Assume $m = \dim M$ is greater than 10, then the following calculations hold provided $1 < n < (m-7)/3$:

$$\pi_n(\mathrm{Top}\,M) \otimes \mathbb{Z}[\frac{1}{\nu}] = 0, \quad and$$

$$\pi_n(\mathrm{Diff}\,M) \otimes \mathbb{Q} = \begin{cases} \oplus_{j=1}^{\infty} H_{(s+1)-4j}(M, \mathbb{Q}), & \text{if } m \text{ is odd} \\ 0, & \text{if } m \text{ is even.} \end{cases}$$

We now turn to a discussion of some recent constructions of non-classical closed aspherical manifolds. It had been conjectured that the total space of the universal cover of a closed aspherical manifold must be homeomorphic to Euclidean space. All the classical examples mentioned in Chapter 1 satisfy this conjecture. But Davis [1983] constructed, for each dimension $n \geq 4$, n-dimensional closed aspherical manifolds N such that the total space of the universal cover of N is not homeomorphic to $\mathbb{R}^n$.

It also seemed reasonable that a closed aspherical manifold should support a smooth structure. But Davis and Hausmann [1989] give examples of closed aspherical manifolds which do not support a smooth structure. In fact, Davis and Januszkiewicz [**to appear**] have constructed such examples in every dimension ≥ 4. (Such examples cannot exist in dimensions 1, 2 and 3.) Moreover, they construct a four-dimensional closed aspherical manifold which cannot be triangulated; i.e., it is not homeomorphic to the geometric realization of a finite simplicial complex.

It had also been conjectured that every strictly negatively curved, closed Riemannian manifold is diffeomorphic to a locally symmetric space. Mostow and Siu [**1980**] gave a four-dimensional counterexample to this conjecture. Gromov and Thurston [**1987**] gave counterexamples in every dimension > 3.

Let N_1 and N_2 be a pair of closed strictly negatively curved manifolds with isomorphic fundamental groups. By results of Eells and Sampson [**1964**], Hartman [**1967**] and Al'ber [**1968**], there exists a unique harmonic map $f: N_1 \to N_2$ inducing this isomorphism. Since N_1 and N_2 are aspherical, f is a homotopy equivalence. When N_1 and N_2 are hyperbolic (and $\dim N_1 > 2$) 1.25 implies that f is an isometry. This led Lawson and Yau to conjecture that f is always a diffeomorphism. Farrell and Jones [**1989c**] gave the first counterexample to this conjecture. It is a consequence of the following result.

6.5. THEOREM (Farrell and Jones [**1989c**]). *Let N be a closed hyperbolic manifold with $\dim N > 4$. Given any $\epsilon > 0$, there exists a finite sheeted cover $\mathcal{N}$ of N such that the following is true.*

(1) *The connected sum $\mathcal{N}\#\Sigma$, where Σ is any closed smooth manifold homotopically equivalent to S^m, supports a Riemannian metric such that all its sectional curvatures are pinched within ϵ of -1.*

(2) *Let Σ_1 and Σ_2 be any nondiffeomorphic pair of closed smooth manifolds which are both homotopically equivalent to S^m. Then $\mathcal{N}\#\Sigma_1$ is not diffeomorphic to $\mathcal{N}\#\Sigma_2$.*

COUNTEREXAMPLE TO LAWSON–YAU CONJECTURE. Pick a closed smooth manifold Σ such that Σ is homeomorphic to S^m ($m > 4$) but not diffeomorphic to S^m. Milnor [**1956**] and Kervaire–Milnor [**1963**] give many such Σ. Borel's Theorem 4.5 yields the existence a closed hyperbolic m-dimensional manifold N^m. Let ϵ be any positive number less than 1 and $\mathcal{N}$ be the covering space of N posited in 6.5. Then $N_1 = \mathcal{N}$ and $N_2 = \mathcal{N}\#\Sigma$, where $\mathcal{N}$ is given the hyperbolic metric induced from N and $\mathcal{N}\#\Sigma$ the metric posited in (1) of 6.5.

REMARK. A partial positive result related to the Lawson–Yau conjecture is a consequence of 6.1 (cf. Farrell and Jones [**1989 b**]). Namely, there is always a homeomorphism inducing the isomorphism between $\pi_1 N_1$ and $\pi_1 N_2$ provided $\dim N_1 \neq 3, 4$.

References

S. I. Al'ber [1968], *Spaces of mappings into manifold of negative curvature*, Dokl. Akad. Nauk USSR **178** (1968), 13–16.

J. W. Alexander [1923], *On the deformation of an n-cell*, Proc. Nat. Acad. Sci. U.S.A. **9** (1923), 406–407.

D. R. Anderson [1972], *The Whitehead torsion of the total space of a fiber bundle*, Topology **11** (1972), 179–194.

B. A. Artamonov [1981], *Projective, nonfree modules over group rings of solvable groups*, (Russian) Mat. sb. (N.S.) **116** (1981), 232–244.

M. Atiyah [1967], *K-Theory*, Benjamin, New York, 1967.

L. Auslander and F. E. A. Johnson [1976], *On a conjecture of C. T. C. Wall*, J. London Math. Soc. (2) **14** (1976), 331–332.

W. Ballman, M. Gromov and V. Schroeder [1985], *Manifolds of nonpositive curvature*, Birkhäuser, Boston, 1985.

D. Barden [1963], *The structure of manifolds*, Ph.D. Thesis, Cambridge University, Cambridge, England, 1963.

H. Bass, A. Heller and R. Swan [1964], *The Whitehead group of a polynomial extension*, Inst. Hautes Études Sci. Publ. Math. **22** (1964), 64–79.

H. Bass and M. P. Murthy [1967], *Grothendieck groups and Picard groups of abelian group rings*, Ann. of Math. **86** (1967), 16–73.

P. H. Berridge and M. J. Dunwoody [1979], *Non-free projective modules for torsion-free groups*, J. London Math. Soc. (2) **19** (1979), 433–436.

L. Bieberbach [1910], *Über die Bewegungsgruppen des n-dimensionalen euklidischen Raumes mit einem endlichen Fundamentalbereich*, Gött Nachr. (1910), 75–84.

L. Bieberbach [1912], *Über die Bewegungsgruppen der Euklidischen Räume II, Die Gruppen mit einem endlichen Fundamentalbereich*, Math. Ann. **72** (1912), 400–412.

A. Borel [1963], *Compact Clifford-Klein forms of symmetric spaces*, Topology **2** (1963), 111–112.

N. Boyom [preprint 1984], *The lifting problem for affine structures in nilpotent Lie groups* (to appear).

W. Browder [1962], *Homotopy type of differentiable manifolds*, Colloq. on Alg. Top. Notes, Aarhus (1962), 42–46.

W. Browder [1965], *On the action of $\theta^n(\partial\pi)$*, Differential and Combinatorial Topology, Princeton University Press, Princeton, N.J., 1965, pp. 23–36.

S. Cappell [1976], *A splitting theorem for manifolds*, Invent. Math. **33** (1976), 69–170.

D. W. Carter [1980], *Localization in lower algebraic K-theory*, Comm. Algebra **8** (1980), 603–622.

T. A. Chapman [1974], *Topological invariance of Whitehead torsion*, Amer. J. Math. **93** (1974), 488–497.

T. A. Chapman and S. Ferry [1979], *Approximating homotopy equivalences by homeomorphism*, American J. Math. **101** (1979), 567–582.

M. Cohen [1973], *A course in simple-homotopy theory*, Springer Graduate Texts in Math. **10**, New York, 1973.

M. Davis [1983], *Groups generated by reflections and aspherical manifolds not covered by Euclidean space*, Annals of Math. **117** (1983), 293–325.

M. Davis and J.-C. Hausmann [1989], *Aspherical manifolds without smooth or PL structure*, Lecture Notes in Math., vol. 1370, Springer-Verlag, New York, 1989, pp. 135–142.

M. Davis and T. Januszkiewicz, *Hyperbolization of polyhedra, J. Differential Geom., to appear.*

S. Donaldson [1983], *An application of gauge theory to four dimensional topology*, J. Differential Geom. **18** (1983), 279–315.

P. Eberlein and B. O'Neill [1973], *Visibility manifolds*, Pacific J. Math. **46** (1973), 45–109.

J. Eells and J. H. Sampson [1964], *Harmonic mappings of Riemannian manifolds*, Amer. J. Math. **86** (1964), 109–160.

D. Epstein and M. Shub [1968], *Expanding endomorphism of flat manifolds*, Topology **7** (1968), 139–141.

F. T. Farrell [1971], *The obstruction to fibering a manifold over a circle*, Indiana University Math. J. **21** (1971), 315–346.

F. T. Farrell [1977], *The non-finiteness of* Nil, Proc. Amer. Math. Soc. **65** (1977), 215–216.

F. T. Farrell and W. C. Hsiang [1970], *A formula for* $K_1(R_\alpha[T])$, Proc. Symp. Pure Math. v. 17, Applications of Categorical Algebra, American Mathematical Society (1970), Providence, R.I., pp. 192–218.

F. T. Farrell and W. C. Hsiang [1978], *The topological-Euclidean space form problem*, Invent. Math. **45** (1978), 181–192.

F. T. Farrell and W. C. Hsiang [1978b], *On the rational homotopy groups of the diffeomorphism groups of discs, spheres and aspherical manifolds*, Proc. Sympos. Pure Math., Vol. 32, American Math. Soc., Providence, R.I., 1978, pp. 325–337.

F. T. Farrell and W. C. Hsiang [1981], *The Whitehead group of poly-(finite or cyclic) groups*, J. London Math. Soc. (2) **24** (1981), 308–324.

F. T. Farrell and W. C. Hsiang [1981b], *On Novikov's conjecture for nonpositively curved manifolds*, I, Ann. of Math. (2) **113** (1981), 199–209.

F. T. Farrell and W. C. Hsiang [1983], *Topological characterization of flat and almost flat Riemannian manifolds* $M^n (n \neq 3, 4)$, Amer. J. Math. **105** (1983), 641–672.

F. T. Farrell and W. C. Hsiang [1983b], *A structure set analogue of Champman-Ferry-Quinn theory*, Continua, Decomposition, Manifolds, (R. H. Bing, W. T. Eaton, M. P. Starbird (eds.)), University of Texas Press, Austin, 1983.

F. T. Farrell and W. C. Hsiang [1984], *On Novikov's conjecture for cocompact discrete subgroups of a Lie group*, Lecture Notes in Math., Springer, Berlin, vol 1051, 1984, pp. 38–48.

F. T. Farrell and W. C. Hsiang [1987], *Addendum to: On Novikov's conjecture for cocompact subgroups of a Lie group*, Topology Appl. **26** (1987), 93–95.

F. T. Farrell and L. E. Jones [1986], *K-theory and dynamics* I, Ann. of Math. **124** (1986), 531–569.

F. T. Farrell and L. E. Jones [1986b], *Algebraic K-theory of spaces stratified fibered over hyperbolic orbifolds*, Proc. Nat. Acad. Sci. U.S.A. **83** (1986), 5364–5366.

F. T. Farrell and L. E. Jones [1987], *K-theory and dynamics* II, Ann. of Math **126** (1987), 451–493.

F. T. Farrell and L. E. Jones [1987b], *Algebraic K-theory of discrete subgroups of Lie groups*, Proc. Nat. Acad. Sci. U.S.A. **84** (1987), 3095–3096.

F. T. Farrell and L. E. Jones [1988], *The surgery L-groups of poly-(finite or cyclic) groups*, Invent. Math. **91** (1988), 559–586.

F. T. Farrell and L. E. Jones [1988b], *Foliated control theory* I, K-theory **2** (1988), 357–399.

F. T. Farrell and L. E. Jones [1988c], *Foliated control theory* II, K-theory **2** (1988), 401–430.

F. T. Farrell and L. E. Jones [1989], *A topological analogue of Mostow's rigidity theorem*, J. Amer. Math. Soc. **2** (1989), 257–370.

F. T. Farrell and L. E. Jones [1989b], *Compact negatively curved manifolds (of* dim $\neq 3, 4$) *are topologically rigid*, Proc. Nat. Acad. of Sci. U.S.A., **86** (1989), 3461–3463.

F. T. Farrell and L. E. Jones [1989c], *Negatively curved manifolds with exotic smooth struc-tures*, J. Amer. Math. Soc. **2** (1989), 899–908.

F. T. Farrell and L. E. Jones [1990], *Rigidity and other topological aspects of compact non-posi-tively curved manifolds*, Bull. Amer. Math. Soc. (to appear).

S. Ferry [1977], *The homeomorphism group of a compact Hilbert cube manifold is an ANR*, Ann. of Math. **106** (1977), 101–119.

M. H. Freedman [1982], *The topology of four-dimensional manifolds*, J. Differential Geometry **17** (1982), 357–453.

M. Freedman and F. Quinn [to appear], *Topology of 4-manifolds*, Princeton University Press, Princeton, New Jersey, USA.

M. Gromov [1978], *Almost flat manifolds*, J. Differential Geometry **13** (1978), 231–241.

M. Gromov and W. Thurston [1987], *Pinching constants for hyperbolic manifolds*, Invent. Math. **89** (1987), 1–12.

P. Hartman [1967], *On homotopic harmonic maps*, Canad. J. Math. **19** (1967), 673–687.

A. E. Hatcher [1975], *Higher simple homotopy theory*, Ann. of Math. **102** (1975), 101–137.

A. E. Hatcher [1978], *Concordance spaces, higher simple homotopy theory, and applications*, Proc. Sympos. Pure Math., Vol. 32, Amer. Math. Soc., Providence, R.I., 1978, pp. 3–21.

A. Hatcher and J. Wagoner [1973], *Pseudo-isotopies of compact manifolds*, Astérisque **6** (1973).

N. J. Hicks [1971], *Notes on Differential Geometry*, Van Nostrand Reinhold Co., London, 1971.

G. Higman [1940], *The units of groups rings*, Proc. London Math. Soc. **46**, 231–248.

M. Hirsch [1959], *Immersions of manifolds*, Trans. Amer. Math. Soc. **93** (1959), 242–276.

F. Hirzebruch [1966], *Topological Methods in Algebraic Geometry*, third edition, Springer-Verlag, New York, 1966.

W. C. Hsiang [1969], *A splitting theorem and the Künneth formula in algebraic K-theory*, Lecture Notes in Math. vol. 108, Springer-Verlag, Berlin and New York, 1969, pp. 72–77.

K. Igusa [1988], *The stability theorem for pseudoisotopies*, K-theory **2** (1988), 1–355.

G. G. Kasparov [1970], *The homotopy invariance of rational Pontrjagin numbers*, Dokl. Akad. Nauk SSSR **190** (1970), 1022–1025.

G. G. Kasparov [1988], *Equivariant KK-theory and the Novikov conjecture*, Invent. Math. **91** (1988), 147–201.

M. Kervaire and J. W. Milnor [1963], *Groups of homotopy spheres: I*, Ann. of Math. **77** (1963), 504–537.

R. C. Kirby and L. C. Siebenmann [1977], *Foundational Essays on Topological Manifolds, Smoothings, and Triangulations*, Annals of Math. Studies, Princeton University Press, Princeton, 1977.

J.-L. Loday [1976], *K-théorie algébrique et représentations de groupes*, Ann. Sci. École Norm. Sup. **9** (1976), 309–377.

A. I. Malcev [1951], *On a class of homogeneous spaces*, Amer. Math. Soc. Transl. No. 39 (1951).

G. A. Margulis [1977], *Discrete group of motions of manifolds of nonpositive curvature*, Amer. Math. Soc. Transl. (2) Vol. 109 (1977), 33–45.

G. A. Margulis [1983], *Free totally discontinuous groups of affine transformations*, Soviet Math. Dokl. **28** (1983), 435–439.

B. Mazur [1963], *Relative neighborhoods and the theorems of Smale*, Ann. of Math. **77** (1963), 232–249.

J. W. Milnor [1958], *On the existence of a connection with curvature zero*, Comment. Math. Helv. **32** (1958), 215–223.

J. W. Milnor [1956], *On manifolds homeomorphic to the 7-sphere*, Ann. of Math. **64** (1956), 399–405.

J. W. Milnor [1962], *A unique factorization theorem for three-manifolds*, Amer. J. Math. **84** (1962), 1–7.

J. W. Milnor [1977], *On fundamental groups of complete affinely flat manifolds*, Adv. in Math. **25** (1977), 178–187.

J. W. Milnor and J. D. Stasheff [1974], *Characteristic Classes*, Princeton University Press, Princeton, 1974.

A. S. Mishchenko [1974], *Infinite dimensional representations of discrete groups and higher signatures*, Izv. Akad. Nauk SSSR Ser. Mat. **38** (1974), 81–106.

E. E. Moise [1952], *Affine structures in three-manifolds V , the triangulation theorem and Hauptvermutung*, Ann. of Math. **55** (1952), 96–114.

G. D. Mostow [1954], *Factors spaces of solvable groups*, Ann. of Math. **60** (1954), 1–27.

G. D. Mostow [1957], *On the fundamental group of homogeneous spaces*, Ann. of Math **66** (1957), 249–255.

G. D. Mostow [1967], *Quasi-conformal mappings in n-space and the rigidity of hyperbolic space forms*, Inst. Hautes Études Sci. Publ. Math. **34** (1967), 53–104.

G. D. Mostow [1973], *Strong Rigidity of Locally Symmetric Spaces*, Princeton Univ. Press, Princeton, 1973.

G. D. Mostow and Y. T. Siu [1980], *A compact Kähler surface of negative curvature not covered by the ball*, Ann. of Math. **112** (1980), 321–360.

M. H. A. Newman [1966], *The engulfing theorem for topological manifolds*, Ann. of Math. **84** (1966), 555–571.

A. J. Nicas [1985], *On the higher Whitehead groups of a Bieberbach group*, Trans. Amer. Math. Soc. **287** (1985), 853–859.

A. J. Nicas [1986], *An infinite family of non-Haken hyperbolic three-manifolds with vanishing Whitehead groups*, Math. Proc. Cambridge Philos. Soc. **99** (1986), 239–246.

A. J. Nicas and C. W. Stark [1984], *Whitehead groups of certain hyperbolic manifolds*, Math. Proc. Cambridge Philos. Soc. **95** (1984), 299–308.

A. J. Nicas and C. W. Stark [1987], *Whitehead groups of certain hyperbolic manifolds* II, Combinatorial Group Theory and Topology, Annals of Math Studies, Vol. III, Princeton University Press, Princeton, N.J., 1987, pp. 415–432.

S. P. Novikov [1964], *Homotopically equivalent smooth manifolds*, Izv. Akad. Nauk SSSR Ser. Mat. **28** (1964), 365–474.

S. P. Novikov [1965], *Homotopic and topological invariance of certain rational classes of Pontryagin*, Doklady **162** (1965), 854–857.

S. P. Novikov [1966], *On manifolds with free abelian fundamental group and their applications*, Izv. Akad. Nauk SSSR **30** (1966), 207–246.

H. Poincaré [1904], *Cinquième complément à l'Analysis situs*, Rend. Circ. Mat. Palermo **18** (1904), 45–110.

D. Quillen [1973], *Higher algebraic K-theory*. I, Algebraic K-theory, I, Lecture Notes in Math, vol. 341, Springer-Verlag, Berlin and New York, 1973, pp. 85–147.

D. Quillen [1976], *Projective modules over polynomial rings*, Invent. Math. **36** (1976), 167–172.

F. Quinn [1970], *A geometric formulation of surgery*, Topology of Manifolds, Markham, Chicago, 1970, pp. 500–511.

F. Quinn [1979], *Ends of maps* I, Ann. of Math. **110** (1979), 275–331.

V. A. Rohlin [1966], *Pontryagin-Hirzebruch class of codimension 2* , Izv. Akad. Nauk SSSR Ser. Mat. **30** (1966), 705–718.

E. Ruh [1982], *Almost flat manifolds*, J. Differential Geometry **17** (1982), 1–14.

S. Smale [1961], *Generalized Poincaré's conjecture in dimensions greater than four*, Ann. of Math. **74** (1961), 391–406.

S. Smale [1962], *On structure of manifolds*, Amer. J. Math. **84** (1962), 387–399.

J. Stallings [1962], *On fibering certain three-manifolds*, Topology of three manifolds, Prentice Hall, Englewood Cliffs, N.J., 1962, pp. 95–100.

J. Stallings [1965], *Infinite processess*, Differential and Combinatorial topology (S. Cairn, ed.), Princeton University Press, Princeton, New Jersey, 1965, pp. 245–253.

A. A. Suslin [1976], *Projective modules over a polynomial ring are free* (Russian), Dokl. Akad. Nauk. SSSR **229** (1976), 1063–1066.

R. G. Swan [1970], *K-theory of finite groups and orders*, Lecture Notes in Math, vol 149, Springer-Verlag, Berlin and New York, 1970.

R. G. Swan [1978], *Projective modules over Laurent polynomial rings*, Trans. Amer. Math. Soc. **237** (1978), 111–120.

W. Thurston [1982], *Three dimensional manifolds*, Kleinian groups and hyperbolic geometry, Bull. Amer. Math. Soc. **6** (1982), 357–381.

J. Tits [1972], *Free subgroups in linear groups*, J. Algebra **20** (1972), 250–270.

F. Waldhausen [1973], *Whitehead groups of generalized free products*, Algebraic K-theory II, Lecture Notes in Math, Vol 342, Springer-Verlag, Berlin, 1970, pp. 155–179.

F. Waldhausen [1978], *Algebraic K-theory of topological spaces* I, Proc. Symp. Pure Math. **32** (1978), 35–60.

F. Waldhausen [1978b], *Algebraic K-theory of generalized free products*, Ann. of Math. **108** (1978), 135–256.

C. T. C. Wall [1971], *Surgery on compact manifolds*, Academic Press, London, 1971.

J. H. C. Whitehead [1939], *Simplicial spaces, nucleii and m-groups*, Proc. London Math. Soc. **45** (1939), 243–327.

H. Zassenhaus [1948], *Über einen Algorithmus Bestimmung der Raumgruppen*, Comment. Math. Helv. **21** (1948), 117–141.